Autumn

AUTUMN

A Season of Change

Text and Photographs by

PETER J. MARCHAND

University Press of New England

Hanover and London

University Press of New England, Hanover, NH 03755

© 2000 by University Press of New England

Printed in the United States of America

5   4   3   2   1

Library of Congress Cataloging-in-Publication Data

Marchand, Peter J.

   Autumn : a season of change / Peter J. Marchand.

   p.   cm

   Includes bibliographical references (p.   )

   ISBN 0-87451-869-5 (cloth : alk. paper)—ISBN 0-87451-870-9 (pbk. : alk. paper)

   1. Nature. 2. Science. 3. Autumn. I. Title.

   QH81 .M2653 2000

   508.2—dc21   99-54701

For Vera . . .

for enduring friendship and good times
kickin' leaves in autumn woods

Contents

Illustrations follow pages 18, 50, 82, and 114. Captions for photographs appear on the page following each group.

Preface

When I first proposed this work, some concern was expressed about the practicality of a book on autumn—about its being too narrow in scope, the season itself too fleeting to be of more than a few weeks' interest. But autumn, if the history of American and European literature is any indication, has always been a time for celebrating life, a time when we feel especially invigorated, a time when nature seems most benign and appealing and her gifts most bountiful. John Kieran captured these feelings a half century ago as he recalled a September day spent riding his sorrel across fields and "down the colored archways of country lanes." At twilight, the maples "seemed to be lighted from within," and he recounted, "I felt the beauty of such surroundings to whatever depths were in me, and I still recall the very spot on the wood road where, in the thrill of one of those Autumn rides, I said to myself: 'This is surpassingly beautiful and I love it. Never in the future can I be any happier than I am now.'"[1]

If ever a season commanded attention, even of the unmindful, autumn is it. Cornelius Weygandt observed 65 years ago that when the leaves turn, "prosaic folks even, . . . plodders, stay-at-homes, . . . feel the call to be abroad, to wander, to hunt, to know again the joy of muscles equal to the unusual demands of their directing will."[2] And his observation appears as true now as it was then, for I have never met a person who is ambivalent about the season. The changes occurring at this time of year seem to tug at something within all of us—something, perhaps, that ties to our remotest experience as a species.

This book is about both the beauty and the challenges of the season; about a time of plenty, but also a time of impending stress when plants and animals—often including humans—must make profound adjustments for coming events that many have never before experienced. The book has as its central theme the many biological nuances particular to autumn, but the subject is life itself. And while the book is primarily concerned with nature and science, it is also about art, literature, and a smattering of philosophy, because to discuss science in isolation of these would be something like studying the life of a flying squirrel in a cage: fruitful to a point, but missing essential context. I have always believed that science and art are natural partners, for both are a way of seeing, both abstractions of the world around us. While one discipline is more cognitive, the other more affectual, both seek to portray the realities of life, each in search of its own truth. Science (perhaps more than art) also builds on history, and history is embodied in the literature of our predecessors who saw things in the light of very different experiences than our own. And insofar as science dwells on mystery, it also embraces a certain amount of personal philosohy. Science is not as objective as we would like to believe.

Nineteenth-century writer John Burroughs had a special knack for capturing the most sentient elements of his natural surroundings on paper. His talent had much to do, I suspect, with his sense for the aesthetics of nature, rather than with the minutiae, reporting "only such things as impressed me and stuck to me and tasted good." It was often the fresh, rather than the familiar, that he chose to write about—a philosophical preference best expressed through an anecdote he related in his own preface to *Winter Sunshine*:

> I have been told that De Lolme, who wrote a notable book on the English Constitution, said that after he had been in England a few weeks, he fully made up his mind to write a book on that country; after he had lived there a year, he still thought of writing a book, but was not so certain about it, but that after a residence of ten years he abandoned his first design altogether. Instead of furnishing an argument against writing out one's first impressions of a country, I think the experience of the Frenchman shows the importance of doing it at once. The sensations of the first day are what we want—the first flush of the traveler's thought and feeling, before his perception and sensibilities become cloyed or blunted, or before he in any way becomes a part of that which he would observe and describe.[3]

It was a philosophy that obviously suited Burroughs well, for with it he produced some of the most colorful, most beautifully descriptive nature writing of his time.

For me, however, it is far too late. I have been involved in the study of natural science too long to avoid becoming "a part of that which [I] would observe and describe"; to avoid seeing in large measure that which I have been trained to see. This is not all bad, for if the minutia sometimes distracts from the sentient, it is equally capable of deepening the inspiration, and in gaining some understanding of the workings of nature, the pleasures of sight, sound, and smell become even more intensely satisfying. Still, when I am sometimes engrossed in thought about, say, autumn haze and the chemistry of organic aerosols from decaying leaves, Burroughs's description of the October air as "tart and pungent, like the flavor of red-cheeked apples by the roadside,"[4] has an appealing freshness and uncluttered truth about it.

I have decided, therefore, to blend the (relatively) unfiltered observations of the early naturalists, whose words alone bring color to a black-and-white page, with sufficient science to elevate our sense of wonder at the nature around us. It was an arrogant decision, elbowing my way into co-authorship with such greats as Burroughs, Richard Jefferies, Martha McCulloch Williams, Thoreau, and others who can no longer refuse me, but their contribution adds an important sense of history, as well as that "first flush of thought and feelings" (more so than my own, at any rate) that Burroughs admonished on us. Their writing has also presented a serious challenge to my own. Humbling it has been to discover that in many cases a hundred-year advantage in science has allowed me to do little more than slightly embellish the observations of these exemplary nineteenth- and early twentieth-century naturalist-writers.

I have elected in this work to limit my discussion of autumn primarily to terrestrial environments—mostly wooded—at middle and northern latitudes, for it is only in these areas that I can bring personal experience to my writing. I have not attempted comprehensive coverage of all plants and animals within this area but rather have tried to touch upon a broad range of phenomena that typify the kinds of activities and preparations associated with the season. This bias will leave some unsatisfied, for there is much—no less fascinating than the subjects I have chosen to write about—going on in the great oceans, estuaries, and rivers of the world at this time of year. For these omissions, I must ask forgiveness. If through this shortcoming I inspire someone else to continue where I left off, I would be delighted.

Finally, I have sought accuracy in my own depictions of the season through the camera lens, attempting to record details as they are and not as I have come to view them. It was my hope also that black-and-white photography would help to see beyond the sometimes distracting color of the season—and to some extent I believe it has worked. At least it seems to have sharpened my own vision. But alas, bias still creeps in. The camera is a great editor, too, isolating and freezing moments out of context. And I, after all, chose those moments and selected the subject. If in so doing I have failed science in any way by my subjectivity, I'll suffer my arrogance and declare it art.

May you enjoy your world all the more for this small effort.

September 1999   P.J.M.

Acknowledgments

My deepest gratitude goes to the entire staff of University Press of New England for their unfailing support on this and other projects over the years; to my friend Vera Walters who waded patiently through several drafts of the present work, helping me to shape it into an acceptable manuscript; and to my colleagues Joseph Merritt, director of Carnegie Museum's Powdermill Biological Station, and Charles Johnson, Vermont state naturalist, who kept my science honest. Any success I might claim as a writer is due in no small part to their help.

The task of collecting material for this book was made easier by the generosity of many friends along the way who opened their houses to me, even fed me, while I luxuriated in autumn color under the excuse of working on this project. Peter and Shirley Opstrup, Kathy Hutchins, and my family in New England, along with Marc and Barbara Snyder, Alex Vargo, and Sara Stevens in Colorado, are deserving of particular mention. I am equally indebted to Dick Storey, chairman of the Biology Department at Colorado College, for his generous institutional support of this endeavor, and to Audrey Benedict of Cloud Ridge Naturalists for giving me the title to this book and the opportunity to field test some of my material before going to print.

To the truly exceptional people at Gerard's Photo Lab in Colorado Springs and Photographic Works in Tucson, I offer my sincerest thanks for patience beyond call and for extraordinary skill in the darkroom.

I.   A SEASON OF CHANGE

Keepers of Time

❧ One mid-August morning in the high desert of Arizona, with the promise of monsoon rains rumbling in distant clouds and the metallic buzz of cicadas portending another hot summer day, I stepped out of my rock house on top of the mesa and immediately registered a sensation of fall. It took me quite by surprise, like a rustling in the bushes, and I looked around expecting to see what it was that struck me, as if "fall" were an object of some kind. There was nothing, of course; but the sensation was real and lingered like early morning frost in a deep shade.

It was a fleeting incident, one I might have shrugged off without further contemplation, except that something kept drawing me back to it, kept puzzling me as to why I should have experienced it at all. By the calendar, autumn was undoubtedly progressing in the north, but why, I wondered—without any conscious awareness and indeed after many years removed from my native forest environment—was my sense of timing still so closely synchronized with events I could not see? Was there some biological value to humans in monitoring seasonal progression, a vestige of our mammalian heritage, a carryover from an ancient ancestry in which an ability to anticipate change well in advance of its arrival may have been of adaptive value, especially in the face of winter's uncertainty? And what could I have been responding to on that clear desert morning? Were there cells in my brain coded to measure day length and send out hormonal messages that say "pay attention, things are changing"? To such questions, Edward O. Wilson, the eminent naturalist and sociobiologist from Harvard, would likely have answered yes, suggesting that with only a few thousand years of separation we might well carry genes from ancestral hunter-gatherers whose fitness might have increased with the sensitivity of their own biological calendars.

If this were so, I wondered, could such a link to our distant past account for other emotions that seem deeply rooted within my core—a need, for example, to be roaming the woods at this time of year; an urge to pick up the gun and go in search of game, though I am not a hunter?

Others have felt it too: "the time of the chase, the season of the buck and doe and of the ripening of all forest fruits; the time when all men are incipient hunters."[1] Even Thoreau, one of the most sensitive observers of natural history, who hunted as a boy but had long since shunned meat and given up the gun even for his bird studies, admitted to such feelings:

> I found myself ranging the woods, like a half-starved hound, with a strange abandonment, seeking some kind of venison which I might devour, and no morsel could have been too savage for me. The wildest scenes had become unaccountably familiar. I found in myself, and still find, an instinct toward a higher ... spiritual life, as do most men, and another toward a primitive rank and savage one, and I reverence them both. I love the wild not less than the good ... even in civilized communities, the embryo man passes through the hunter stage of development.[2]

Perhaps these responses—the sense of timing, the urges within—simply are conditioned by experience and long habit. Yet if such learned behavior were to be of any benefit to ourselves or our ancestors in preparing for difficult times ahead, it is implicit that we have the ability to anticipate change, which requires measuring something and then comparing this to yesterday's measurement, or last week's, or last month's, which also requires memory in one form or another.

That we might possess the ability to carry out these computations subconsciously does not tax the imagination excessively, but the problem gets much more difficult when we consider an animal of temperate or northern latitudes, born in the spring, whose first winter's survival now requires numerous timely adaptations in physiology and behavior. Less than 6 months old, it must register impending change and make profound adjustments for a season it does not know. Plants, too, must have a way of anticipating change, for in acclimating to winter conditions, timing is critical. Survival of freezing, for example, requires many alterations at the cellular level that must be completed well in advance of the immediate need.

Virtually every organism, then, native to temperate or higher latitudes—and that lives for more than one growing season—must possess a physiological yardstick or calendar, a means of tracking time either directly through internal rhythms or indirectly by monitoring changes in some tangible quality of the environment. What might provide such a precise clock or calendar? Understanding the possibilities first requires understanding the nature of seasonal change.

. . .

Autumn, though more subtle in some areas than others, is universal outside of the tropics. When a newcomer to an area such as the short-grass prairie that generally lacks colorful deciduous trees remarks that he or she misses the seasons, what the person usually means is that he or she misses the turning leaves of aspens, maples, birches, or some other conspicuous indicator of the season. But the changing of leaves does not bring on the season. The absence of deciduous trees in the shortgrass prairie does not mean that fall, any more than spring, fails to come to the grasslands. It is the steady, progressive change in angle of incoming sunlight as the earth, tilted on its axis, voyages around the sun that determines the annual march of seasons; and it is only the intensity of change, increasing with distance from equatorial regions, and the vegetative expression of seasons that varies geographically.

The earth in its voyage around the sun traces an imaginary path that defines the plane of our orbit in space. The sun does not sit at the center of the plane, however, for our orbit is elliptical, with the earth passing closest to the sun in the latter part of December and reaching its most distant point in June. It is not, therefore, our elliptical orbit, our varying distance from the sun, that results in seasons of warmth or coolness, for if this were so our northern summers would occur from December through February as in the southern hemisphere. Rather, it is the tilt of the earth's own axis from vertical (with respect to the plane of our orbit around the sun) that results in large monthly variations of incoming solar radiation at any given location outside the tropics. As we approach the aphelion of our orbit on the summer solstice, the northern hemisphere is tilted prominently toward the sun, giving all the northern latitudes longer, warmer days and keeping the region above the Arctic Circle entirely within the sunlit sector as the earth turns. Six months later, we (in the northern hemisphere) are closer to the sun but tilted away from it. Between these two turning points in our annual journey are the seasons of spring and fall.

Transitions between summer and winter are seldom as smooth, however, as described by planetary physics. Of spring and fall Burroughs remarked, "the [season] comes like a tide running against a strong wind; it is ever beaten back, but ever gaining ground, with now and then a mad push upon the land."[3] Indeed, many of us live in places where spring is occasionally turned on its heels with a surprise May snowstorm, or the bite of autumn frost is temporarily soothed with a placid "Indian summer" of unseasonably warm weather (see p. 122). Though the annual course of incoming solar radiation is entirely predictable, it gives rise to much less dependable changes, on a daily or weekly basis, in the surface temperatures that drive our weather systems. This unpredictability of temperature results from considerable variation over time and distance in the amount of radiant energy absorbed by the earth, as affected by cloudiness, the presence of snow and ice cover, and differences in vegetation or soil character. Thus, temperature alone is too uncertain to serve as the primary signal for the myriad changes that must take place in the fall. The environmental cues by which plants and animals coordinate their seasonal rhythms must be far more dependable.

Three direct aspects of incoming solar radiation, apart from its varying effect on surface temperatures, offer potentially useful measures in predicting the seasons. Each of these is related to the ever-changing angle at which sunlight strikes the earth, as the earth's poles alternately dip toward and then away from the sun. One obvious result of a lower angle of incidence is an increase in shadow length. In order for changes in shadow length to be used effectively as a calendar, however, they would have to be referenced to some fixed measure and standardized for time of day—a practice that was used successfully by prehistoric cultures to anticipate important events, but that would require extraordinary neural function on the part of animals to accomplish subconsciously. More useful, perhaps, is the change in light quality—the spectral character of light reaching us—that accompanies the seasonally changing angle of incidence. Light passing through the atmosphere interacts with various gasses and particulates, each of which tends to absorb or reflect differently, depending on the wavelength or energy of the light. As the distance a beam of sunlight must travel through the earth's atmosphere increases, more attenuation occurs in the ultraviolet and infrared portions of the spectrum, energy just beyond the two ends of the visible spectrum. While this is light the mammalian eye can't see, both ultraviolet and infrared wavelengths are, nonetheless, capable of regulating the biological clocks in rodents (see later discussion). In addition, some attenuation occurs in the blue-green region of the spectrum, which corresponds to the wavelength of peak sensitivity in the biological rhythms of many mammals.[4] Insofar as light quality also changes on a daily basis, however, as between early morning, noon, and evening (hence the prevalence of reds, yellows, and oranges when the sun is near the horizon), some means of time correction would be required if this were to serve as an effective calendar.

The one seasonal trend in solar radiation that seems to affect humans most in fall is the accelerated rate of decrease in total daylight hours or night length. This is particularly dramatic at higher latitudes, of course, where the transition from long summer days ("days" that last for weeks above the Arctic Circle) to equally long winter nights is especially rapid. Day length at 65° N latitude (Fairbanks, Alaska, for example) shortens by 3½ hours between August 1 and September 1. But even at mid latitudes the changing day length at this time of year is easily perceptible, and it is not difficult to imagine this as furnishing a useful calendar, provided some mechanism exists by which an organism can keep track of total hours of sunlight or darkness. The problem seems simple enough for us, as we subconsciously register the slowly changing light conditions of morning and evening relative to clock time and our daily habits of working and eating, but we must keep in mind that for plants and animals the only clocks are internal, and environmental information of this nature must not only be quantified, but also translated into a physiological response.

. . .

If light is to serve as a clock or calendar, something has to receive information about it and translate that information into action. The eye, of course, gathers light, but there is compelling information that it is not the only organ to do so. A congenitally blind person may still show chemical responses to light despite absence of any pupillary reflex, outer retinal functioning, or conscious awareness of a light stimulus. Retinally degenerate mice show normal biological rhythms under simulated day and night light regimes in spite of near total loss of visual perception.[5] And in both birds and insects, the primary receptor of environmental signals regulating reproduction, migration, and fat deposition lies somewhere other than in the retina. Experimentally covering portions of a bird's skull, but not its eyes, can make a vast difference in its developmental response to simulated long or short days. The light receptor in birds and insects, for nonvisual information, appears to lie in the brain, not the eyes.[6]

While the exact location and nature of the extraretinal light sensor in birds and insects remains elusive, in mammals the pineal appears to be the "third eye." The pineal is a single, somewhat club-shaped organ usually found in the midline of the brain at the point where the cerebral hemispheres and the cerebellum come together. It is a distinct organ, separate from the brain, and is highly vascularized—a condition linked to its ultimate function.

The pineal exhibits several properties that suggest a central role as a biological calendar. For one, it is photoreceptive by itself, meaning that it is capable of detecting and responding to light, both electrically and metabolically, even when surgically isolated from the body. The pineal of many animals contains cells with a photosensitive pigment, probably rhodopsin (the same pigment that receives light in the rod cells of the retina). But the pineal also functions in conjunction with neural signals from the brain or nervous system, and is thus capable of integrating light information from the retina as well. Most importantly, the pineal acts like an endocrine gland, a ductless gland that, in response to nerve signals, secretes chemical products into the blood—in this case melatonin, a hormone that regulates a number of developmental and reproductive processes.[7]

In essence, then, while the pineal shows independent light sensitivity, its function as a principal light-monitoring organ in mammals seems tied directly to both the retina and a biological clock that is located in the hypothalamus of the brain. Light impulses from the retina are sent via a special nerve tract directly to the hypothalamus, which coordinates general biological rhythms with day-night cycles. The light information is then relayed by neural signals to the pineal, where it is translated into chemical information via the secretion of melatonin and sent to other parts of the body.[8]

It is the pineal's sensitivity to light information received from the brain and its ability to translate this into chemical information (its endocrine function) that places it in a central position with regard to time measurement. The pineal is essentially the endpoint of an optic system for processing day-length information, not unlike that of the visual cortex in the brain for processing images. The usefulness of the pineal as a calendar, however, hinges on the importance of melatonin as a chemical messenger and its sensitivity to day-night duration. Melatonin is a key player in mediating a number of seasonal changes in animals, many of which are critical to its overwintering success. Introduced artificially into the blood of white-footed mice, melatonin can induce fall molt, even during long days. It can stimulate the accumulation of brown adipose tissue, increase the animals' basal metabolic rate, and double their number of spontaneous daily torpor bouts. Injected into weasels and djungarian hampsters it can prevent molting from white to brown during the lengthening days of spring. Melatonin is also a key hormone in regulating reproductive development in animals. In every case, whether an animal be diurnal or nocturnal in habit, melatonin is synthesized at night, and therefore its quantity in the bloodstream provides direct information to other tissues regarding day length.[9]

A single enzyme is responsible for the light sensitivity of melatonin production. NAT, as it is called (for *N*-acetyltransferase), one of 59 enzymes present in the pineal, catalyzes the conversion of the chemical serotonin to an intermediate product, which is then acted upon by another light-insensitive enzyme to complete the production of melatonin. NAT itself is so sensitive to light that its concentration in the pineal can increase more than 20-fold during a 12-hour night. Longer dark periods result in extended periods of high NAT activity, and hence melatonin secretion, thereby converting light information into a biochemical message in the form of nocturnal melatonin peaks of variable length. Long-duration melatonin peaks signal short days. (Continuous darkness, however, does not continuously stimulate NAT activity and melatonin production. The biological clock functioning in the hypothalamus restricts this activity to periods falling within the "expected" nighttime.) In essence, then, it is night length, rather than day length, that is being measured. If animals are exposed experimentally to a strong light pulse in the middle of the night, NAT levels plummet rapidly, halving in 3 to 5 minutes, with a subsequent drop in melatonin production signaling the end of the dark period. The light pulse is read as a new dawn, and the animal responds physiologically as if the long days of summer had returned.[10]

. . .

If the sensitivity of the mammalian clockworks and calendar seems extraordinary, the timekeeping mechanism of plants is no less so. Lacking a central nervous system to process light impulses from optic sensors, and lacking the integrative function and information distribution capability of an endocrine gland, plants nonetheless anticipate the coming season unfailingly, and do so by acting on the same light information that animals receive. That two such disparate organisms might evolve similar strategies should not be so surprising; the two kingdoms have, after all, responded to the same need, with day length generally the only completely reliable environmental cue available to either. And when it comes to monitoring light, few organisms are better suited to the challenge than green plants, for if photoreceptors depend universally on light-sensitive pigments, then plants have no equal.

Photosynthetic plants have developed an impressive array of pigments designed to capture energy from almost the entire spectrum of visible sunlight. Chlorophyll molecules are the most abundant pigment in leaves, but their light-gathering effectiveness is limited to a relatively narrow color band at the blue-green and red ends of the spectrum. Because environments may differ in the spectral quality of light available to plants (energy reaching the forest floor, for example, has had much of the blue-green and red light filtered out by chlorophyll in the leaves overhead), additional pigments are often employed to capture light of different wavelengths and transfer its energy to the chlorophyll molecules, which are the only ones to participate directly in the chemical reactions of photosynthesis. The accessory pigments are themselves brilliantly colored—primarily shades of yellow and orange—but are not often seen until fall because they are masked by the sheer abundance of chlorophyll during the growing season.

With many different pigments present in the leaf, it would not be illogical to assume that one or another was somehow involved in the measurement of day length. Yet the identity of the clockwork pigment escaped discovery for a long time. When finally the timekeeper was revealed, it turned out to be a very different pigment indeed—one that seemed almost a contradiction in form and function to those previously known. This pigment wasn't found where the others usually were, yet it was almost everywhere in the plant. Its color wasn't far from that of chlorophyll, yet its appearance depended entirely on how, or when, you looked at it. New and fresh, it was bright blue, but the moment the molecule was illuminated by sunlight, it switched to an olive-green color. Exposed specifically to light at the far-red end of the spectrum, however, it reverted immediately back to blue. The pigment had nothing to do with energy capture for photosynthesis, yet by its absorption of light it seemed to have a decided influence on the timing of such key events as the onset of flowering, senescence, and dormancy.

The pigment is called phytochrome, and among all those light-absorbing (and -reflecting) molecules that excite our senses in the fall, this one, the least noticeable, may in many ways be the most important at this time of year. We now know it to serve as both a calendar and an on/off switch for a number of processes essential to the orderly cessation of growth and onset of winter acclimation in the plant. The exact mechanism by which phytochrome works remains a mystery, but the details of its timekeeping ability have largely been solved, and its split personality seems to be the key.

Phytochrome exists in two distinct and reversible states, each sensitive to light of slightly different color and each having greatly different effects on physiological processes in the plant. As noted already, when phytochrome is synthesized from protein by the plant it is blue in color and has an absorption peak (the color band to which a pigment is most responsive) centered in the orange-red region of the spectrum, specifically at a wavelength of 660 nm. In this state, which we call phytochrome-red (or Pr) for its reaction color, the molecule is essentially biologically inert. In fact, when present in abundance it seems to actually block specific developmental processes in the plant.[11]

When phytochrome is excited by absorption of energy in the red band of sunlight, however, it changes state to a form in which it is biologically active, behaving almost like a hormone to open a gate for a number of developmental processes in the plant. In this state the pigment is olive-green in color and no longer reactive to orange-red light. Instead, it absorbs much more efficiently in a region centered near the far end of red, close to the limit of our color vision (we label this form Pfr for far-red)—but when it does so, strangely, it reverts back to its original state. Phytochrome undergoes this same reversal in darkness, too—or decays altogether[12]—and this, it turns out, may be its most useful property as a timekeeper.

How, then, might phytochrome work as a calendar to regulate seasonal events? Sunlight is constantly converting phytochrome from its initial inhibitory form to the active state. But the reverse is also true because the full spectrum of sunlight contains energy affecting both forms of phytochrome (the absorption ranges of Pr and Pfr overlap slightly). Here's an important twist, though: The efficiency with which Pr absorbs red light is greater than the absorption efficiency of Pfr for far-red light, so sunlight acts more like a red source, with more of the Pr ending up as Pfr during the day than is true of the reverse.[13] The result is establishment of an equilibrium concentration of the two forms of phytochrome in daylight, with the biologically active form dominant.

Length of day, by itself, doesn't change this balance, for sunlight will always drive the opposing reactions at about the same rate; but length of night is another matter. With Pfr undergoing either complete destruction and loss during the night, or chemical reversion to its original, inhibitory form, the length of night will have much to do with how long the active form of phytochrome, Pfr, or the blocking form, Pr, is present in the plant at any given season. And in much the same way that exposure to light in the middle of the night will cause a rapid plummet in NAT and melatonin synthesis in mammals, so too will even a 2- or 3-minute dark interruption signal a new dawn in plants.[14]

Thus, the parallels are nearly complete. Like the pineal in the animal, phytochrome is capable of responding to light information and communicating that information via chemical messengers, behaving much like a hormone to switch critical plant processes on or off. It is almost certain that phytochrome does not act alone (calcium is now suspected as an important accomplice[15]), but it is the master. Concealed by the fall brilliance of other pigments, the unseen phytochrome, to the plant at least, may be the real glory of the season.

.   .   .

By mid-August the gradual increase in the length of night begins to make a difference to plants and animals. There are few outward signs of fall yet, for the days are still warm enough to nurture the progeny of summer, but the solstice is already two months past and the shadows of a year's new growth are testing their reach. In the Arctic, the caribou are entering their period of autumnal reproductive development even before the return of true night, with their biological calendar registering the progressively diminishing light intensity around midnight as shortening days.[16] Some of the grasses are browning at their tips as they begin to move reserves of energy and nutrients to the safety of belowground tissues. Flowers are maturing into seed heads of a different beauty. The signs are subtle at first, but the internal timekeepers are at work sending the first messages of impending cold. There is a quiet revolution starting behind the mask of August calm.

WHAT SAITH SEPTEMBER?

A Fair month, truly—golden fair, spiced with breath of the orchards, the vineyards' winy smell . . .

All the earth lies dry and warm, and palpitant in sunshine. The touch of it is vital. Lie at length here in the pasture, prone on its springy turf, and let the strength of it, the sweetness, the balm of healing, lap your tired soul to the Elysium, sleep—such sleep as comes never within four walls, or to the downiest couch ever fashioned by man's hand. Sleep, and dream not. This the hour of fruition, needs not to borrow charm of such insubstantial stuff. A full world and goodly lies all about. Upland orchards blush red and yellow; lowland, stubble, meadow, corn-field, chant in high, colorful notes a swelling prelude to Nature's harvest-home.

What scent comes out of the corn-land—rare, fine, subtile as breath of elfin flowers? All the russet rustling stretch is steeped in its balm. You drink it in long gasps, and turn away, sighing—it is full, so full, of spring, and dew, and dawn, and hope...

A jocund time this should be. The earth, the fulness thereof, lies smiling peace to a perfect heaven. Yet somehow there creeps in an under-note—a wailing minor of loss and waste. Faint, ah, so faint! you hear it in the singing waters, the full, rich, rustling leaves, the low winds sighing out of the sky to lose them as wafts of balm. Through them September saith to this fair world, "Laugh, dance, lie in the sun; eat, drink, and be merry. To-morrow you must die."

Walk afield [then] every day . . . Whether sun shines, or rain drips, or white frost bites and stings, you should find a liberal education in the hectic beauty of death; not cruel death, but a tender doom, sweet with the glory of full harvest, and spanned with the rainbow of spring resurection.

—MARTHA MCCULLOCH WILLIAMS, 1892[17]

Turning

In autumn's "hectic beauty of death," little is left to chance and circumstance. The entire process of senescence in plants—the ageing, the dying back of nonperennial parts, the color change, and the final discarding of stems and leaves—is highly ordered, involving much more than a passive loss of normal function in the plant. Senescence is an active process in which dormancy and protection, rather than growth and reproduction, become the objects of metabolism. There is, in a sense, as much life as there is death in the browning of meadows and the drying of leaves in autumn.

If we could sit under a maple or aspen in the fall and observe all that is going on within the plant, we would witness a remarkable communication of chemicals and flow of materials: a bustling hubbub of messenger molecules and hormones directing complex metabolic processes; a train of nutrients and waste materials being shuttled into storage compartments; a clatter of metabolic machinery being disassembled. We would witness a diminishing concentration of growth hormones—auxins and cytokinins—turning off the processes of cell division, the production of new DNA, the synthesis of various enzymes. Buds become dormant; chlorophyll production winds down. With a reduction of both photosynthesis and growth, we would see unused stores of sugars shipped to roots, rhizomes, or, in some trees (like our maples), into specialized ray cells in the wood, to flow again in the spring.

Accompanying all this movement would be noticeable changes in the structure of leaf cells and organelles. Chloroplast membranes, where photosynthesis earlier took place, begin to swell, and fatty lipids appear, like tiny droplets of oil. Membranes surrounding other organelles show signs of alteration, too. Ribosomes, where protein is synthesized, break apart and slowly disappear from the chloroplasts as well as from the cell at large. Dictysomes, which are the collecting, packaging, and secretion centers of the cell, start to break down, and so, too, does the cell nucleus. Eventually even the mitochondria, the respiratory engines of the cell, begin to shut down, though their persistence until late in the senescent stage testifies to the continuing need for energy to support the many de-activation and translocation processes now taking place.[1]

As the machinery of photosynthesis and respiration is slowly dismantled, many mineral elements are recovered and recycled internally, particularly those like nitrogen and phosphorus that are in high demand but often found in short supply in the soil. A large fraction of the nitrogen-rich protein synthesized during the growing season is either converted into a soluble form or broken into its basic amino acid building blocks and translocated to stem or root tissues. Similarly, the nitrogen contained in nucleic acids (DNA and RNA) and lipids is moved out of the leaves and into permanent tissues where a portion may be incorporated into new proteins and nucleic or amino acids for storage (there are more than 800 different proteins and several forms of nucleic acids in the plant). Thus we see during senescence both the breakdown and simultaneous synthesis of protein and nucleic acids.[2]

Since much of the phosphorus garnered by plants during the growing season is incorporated into large molecules that make up the double-layered membranes surrounding cells and organelles, the gradual breakdown of these structures in nonpermanent tissues yields recyclable phosphorus, too. Along with nitrogen, phosphorus begins moving out of the leaves of deciduous plants as normal cellular metabolism slows, but unlike nitrogen, it may not simply be going into inactive storage. Increased phosphorus levels in stem and bud tissues following its removal from leaves has been implicated in the process of winter acclimation.[3]

· · ·

All of this is, of course, unobservable to us sitting under the tree. But there is one aspect of leaf senescence that most assuredly does not escape our notice, and that is the color change that accompanies the breakdown of chlorophyll. This is the main attraction, the show stopper. The spectacular display of color in our deciduous woods *is* fall to most people.

> How beautifully the leaves grow old! . . . [They] fairly become luminous, as if they glowed with an inner light. In October a maple tree before your window lights up your room like a great lamp.　　JOHN BURROUGHS, 1924[4]

The source of this "inner light" has already been hinted at in our brief discussion of timekeepers and the role of various pigments in the leaf, but let us now fill in some of the details. We have seen that a number of supplemental pigments (termed accessory pigments) exist to capture light energy not absorbed directly by green chlorophyll molecules, but that these pigments are present in relatively small amounts and are not visible until chlorophyll begins to fade. And we know now that chlorophyll degradation is part of a suite of processes associated with the active shutting down of nonpermanent plant tissues—but that is only part of the story behind the changing colors.

There are many accessory pigments in leaves that serve not only to enhance energy capture, but also to provide a number of protective functions. The carotenoid pigments, for example, in addition to trapping light also reduce the sensitivity of chlorophyll to bleaching—the destructive breakdown of chlorophyll in the presence of light and oxygen (paradoxically, sunlight absorbed by chlorophyll can lead to excited states of oxygen, and these highly reactive states can damage the pigment).[5] There are others, too; pigment groups that are not associated with chlorophyll and do not participate in trapping light energy for photosynthesis, but which have critical functions in flowers, fruits, and leaves.

One such group, known collectively as the flavinoids, includes three different pigments that are similar in structure but different in appearance and role. These are the anthocyanins, flavonols, and flavones. Anthocyanins are the bright blue, purple, and red pigments that often color flowers and fruits, and serve primarily to attract pollinators and seed dispersers. Anthocyanins are also present in leaves in small amounts, where their function is uncertain but may be related in some way to disease resistance. The other two members of this group, the flavonols and flavones, are either creamy-white or colorless, and are also commonly found in flower parts, sometimes associated with anthocyanins. In the leaf the flavonols and flavones apparently play a role in blocking far-ultraviolet light, which is biologically damaging, and at the same time selectively pass blue-green and red wavelengths that are optimal for photosynthesis.[6] During senescence, however, the colorless flavonols undergo a simple but remarkable biochemical transformation and become the real magicians of fall color in many species.

The trick lies in one seemingly mundane biochemical detail. As already noted, the three flavinoid pigments vary only slightly from one another in molecular structure. Anthocyanins, for example, differ from the others primarily in having a sugar molecule attached to their carbon framework. But what a difference a sugar makes! This simple alteration of molecular configuration has a dramatic effect on the reflectance characteristics of the pigment, producing the vivid colors we see in flowers and fruits.[7] If, then, a sugar molecule should bond to a flavonol, the result is the same—and herein lies the secret of many an autumn leaf. When plant respiration shuts down and sugars begin to accumulate in the leaf (for lack of demand or transport out of the leaf), they often attach to flavonol molecules, suddenly transforming the colorless flavonols into brilliant anthocyanin pigments of flaming red or striking purplish hues, fanning the fires already fueled by the glowing orange and yellow carotenoids. It would appear to the affective senses a magical collaboration of the physical and biological, a transformation as glorious as any in nature. Slowly, as if bleeding out of the leaf, the carotenoid pigments appear where there seemed to be none before. Then, just as subtly, some trees and shrubs increase production of anthocyanins until they fairly glow with that "inner light"—a light that is nothing less than sunlight itself, bouncing off the very molecules that gave life to the plant in the summertime.

> The . . . forest has caught the trick of the sunset, and glows at the season's setting with all the glory of the evening's western sky.   CHARLES ABBOTT, 1889[8]

. . .

If all this seems simple, it is not. While the chemical transformations that produce color are rather straightforward, the environmental factors that affect the intensity and timing of color change and leaf drop are often confounding.

In the fall of 1995, for example, the spectacular display of color that typifies the West Elk Wilderness of Colorado never came. The West Elk area is aspen country at its finest, and in most years the late-September mountains are an extraordinary sight, glittering with the gold of a trillion shimmering medallions (the locals claim you need dark glasses just to look at the aspen forests!). But that year, though the crisp and sunny weather seemed perfect, the change didn't occur. The leaves simply turned brown and shriveled on the trees before dropping.

An autumn without color provokes many questions. If day length cues the hormonal changes (through the phytochrome mediator) that bring about senescence, there must be some overriding factor(s) that can hasten or delay the process. Not surprisingly, there are many popular claims regarding the kind of weather necessary to bring out the best color in deciduous trees and shrubs, but there is also a notable scarcity of data on the subject (most research efforts on plant senescence are aimed at agricultural or horticultural species, often those of greatest commercial interest). Many ideas about fall foliage change are still based on anecdotal or circumstantial observation, much of which was provided by the early naturalists who were quite intrigued with the phenomenon. Charles Abbott, a natural historian living in New Jersey in the late 1800s, kept numerous records of the turning of foliage in relation to weather, and it was he who first suggested that early growing season conditions might also affect the timing of leaf fall:

> The condition of the growth of the leaves in spring appears to have much to do with the progress of the autumnal change . . . [T]hree enormous beeches near my home were in full foliage, May 1, 1886, and the rich yellow-green of the growing leaves had wholly disappeared. During the first week in June, the branches of one of these trees having an eastern exposure, produced a copious second growth of leaves. In October following, when the foliage generally had dropped, this second growth still held its place, and did not fall until the middle of November. It came a month later and tarried that much longer. The same phenomenon I have often noticed in connection with the many oaks that abound here.[9]

Abbott expounded on this idea with notes about the importance of adequate soil moisture in the early part of the growing season, and the relative insensitivity of leaf fall to late season drought:

> We see, too, the effect of an early drought during the following autumn, for the leaves fall earlier in the season if they were checked in April by want of moisture; but a long drought, as is now so common in August or September, does not affect the leaves . . . to any noticeable extent so far as their falling is concerned . . . [I]n 1886 . . . there occurred a protracted late summer drought, yet the leaves remained upon the trees longer than usual—a fact not to be ascribed to absence of frost, but to the vigor they received from a superabundance of rain in April and early May, a vigor in nowise checked by the low temperature of August 29, when frost formed in damp situations.[10]

We are aware now that any growth processes dependent upon accumulated sugars or starch reserves can be affected by environmental conditions in some cases extending back into the previous growing season—and this complicates our attempts to explain (and certainly to forecast) year-to-year differences in autumn color considerably. But there are proximal causes, too, for annual variation in color and timing, and it is possible from our understanding of leaf senescence to piece together some of the effects of more immediate weather conditions on the process. It is reasonable to suggest that once active growth has ceased, warm, sunny days that promote continued sugar production (albeit at a much reduced rate), followed by cool nights that slow respiration and

The entire landscape shouts of passing time. Color drains slowly from the high country, collecting and intensifying along the lower stream courses. Every turning leaf and bare branch maps the progress of a season and the preparations of whole communities of unseen organisms for a winter many do not know.

The Lit River, purling and eddying onward, was spotted with recently fallen leaves, some of which were being carried round by eddies. Leaves are now falling all the country over . . . some are being conveyed silently seaward on rivers: [others] concealing the water in swamps, where at length they flat out and sink to the bottom, and we never hear of them again, unless we shall see their impressions on the coal of a future geological period . . . The trees repay the earth with interest for what they have taken from it.

—Henry David Thoreau, 1858

Oak and aspen, San Juan Mountains, Colorado.

Harbinger of autumn to come, the ostrich fern is one of the first plants to turn—and first to wither with the nip of frost (p. 31).

Color often comes earliest to stream valleys as cold air settling in low drainages amplifies the effects of shortening days.

Behind the turning colors is a clockwork pigment we cannot see—a molecular timekeeper of remarkable capability (p. 10).

It is not only the eye that sees. Another light sensor tallies shortening daylight hours, triggering autumn molt in the ptarmigan (p. 7).

Seemingly frail and perishable, fall insect larvae prepare for freezing with an extraordinary arsenal of chemical defenses (p. 33).

Aspen leaves on water.

Swirling eddies carry the energy and nutrients of autumn leaves from the land to aquatic ecosystems.

the transport of materials out of the leaf, will promote more intense color in those trees that produce anthocyanins. This effect can often be seen in trees that are partially shaded, where exposure to sun brings out more color on the sunlight side. On the other hand, persistent cloudiness that reduces photosynthesis during the day, while also favoring plant respiration with warmer nighttime temperatures, would lead to lower sugar production and its more rapid removal, resulting in diminished anthocyanin production and duller colors. Extending this analysis to the whole growing season is a bit risky, but might shed light on occasional anomalies like the 1995 West Elk situation. The most conspicuous feature of the weather records for western Colorado during the "year-without-color" was the extraordinarily warm spring experienced at many locations. Monthly average temperatures ranged from 3 to 10 degrees above long-term norms for February, March, and April, a time when increased respiration rates in trees (before the new leaves appear) would rapidly deplete stored energy reserves from the previous year. The warm spring was followed by a growing season that was considerably wetter and cooler than average, likely maintaining below-normal photosynthetic production. With rainfall still slightly above average, temperatures in August and September climbed again, favoring utilization of sugars as they were manufactured. This would have left little in the leaf with which to produce the usual fall color.

Such explanation still leaves many questions unanswered, however, and cautions us to remain skeptical of our understanding, for if the timing and intensity of color change and leaf drop were a simple matter of weather effects, all trees would respond the same, and we know this not to be the case. Genetic differences are also clearly involved, and nowhere is this better seen than on mountainsides containing numerous clones of aspen (clones are offshoots arising from roots, rather than from seed, often forming dense stands of genetically identical trees). Individual clones stand out conspicuously in the fall by their different stages of color development, each programmed genetically to turn according to its own schedule.

WHAT WORDS SHALL PAINT ITS SPLENDID LANGUORS? A rich tint of russet deepened on the forest top, and seemed to sink day by day deeper into the foliage like a stain; riper and riper it grew, as an apple colours. Broad acres these of the last crop, the crop of leaves; a thousand thousand quarters, the broad earth will be their barn. A warm red lies on the hill-side above the woods, as if the red dawn had stayed there through the day; it is the heath and heather seeds; and higher still, a pale yellow fills the larches. The whole of the great hill glows with colour under the short hours of the October sun; and overhead, where the pine-cones hang, the sky is of the deepest azure. The conflagration of the woods burning luminously crowds into those short hours a brilliance the slow summer does not know.   —RICHARD JEFFERIES, 1900[11]

Each waft of winy air brings fleets of fairy argosies—russet, scarlet, gold, and crimson—to anchor on the breast of earth. With what drifts of them the south wind covers fallow and grass land! All the woods are pathless now—footway, cart track, mill road, alike knee-deep in leaves. The highway, even, broad and beaten though it be, shrinks to a ghostly trail through a fluttering world of color, Here big walnuts overhang it, and overhead you see the blue heavens through lacework of bare black boughs, with the faintest flutter of lingering leaves. A little farther, you tramp through the hickory flat. Is there magic abroad? Have genii or gnomes caught you suddenly into a golden world? There is gold all about you—overhead, underfoot. It must be these lithe, gray-stemmed woodland giants stored all of sunshine in their hearts, and now exhale it through their leaves. In the grayest day here is warmth and splendor—a flame of radiance that makes yet darker the sombre oak-wood. Now, when soft winds sift out of a cloudless sky, what words shall paint its splendid languors, its glory of scent and light and color?   —WILLIAMS, 1892[12]

The whole scene had an indefinable look of being painted, the colour was so abstract and correct, and there was something so sketchy and merely impressional about these distant single trees on the horizon that one was forced to think of it all as of a clever French landscape. For it is rather in nature that we see resemblance to art, than in art to nature; and we say a hundred times, "How like a picture!" for once that we say, "How like the truth!   —R. L. STEVENSON, 1905[13]

We might ask at this point where evergreens fit into the picture. Throughout much of the temperate latitudes, evergreen and deciduous trees are found growing in mixed forests under much the same environmental conditions. Similarly, in the vast boreal forests of Canada, Alaska, and Eurasia, where evergreens are overwhelmingly favored, occasional stands of deciduous birch and aspens are found mixed with the conifers. Clearly both are successful, even under the most difficult of conditions. Why, then, have two very different overwintering strategies evolved in the plant kingdom? Why do not all plants turn color and shed their leaves in autumn?

There are essentially two major considerations to be addressed here, one relating to physical stresses during winter and the other to availability of resources for the production of leaves. The presumed advantage of dropping leaves in the fall is to avoid problems of water stress associated with the long winter drought (during winter most water goes into storage in the snowpack, but even where rain falls, absorption of water from cold soils may be problematical). Because the leaf presents a rather large surface area from which evaporation can occur, dropping leaves reduces demand, keeping water needs in line with the lower supply. But the other side of this issue is that leaves need to be produced again in the spring, and this requires considerable nutrient capital. Where soil resources are limited, reinvesting each year in wholly new photosynthetic machinery may not be affordable.

The evergreen habit, then, seems to have evolved as a means of coping with scant nutrient supplies. Evergreens still discard leaves in the fall, but not all at once. Only the older ones, leaves that have functioned for several years, are "retired"—which means that fewer new leaves need to be produced in the spring. And even though the photosynthetic efficiency of 2-, 3-, or 4-year-old evergreen leaves is considerably less than that of new ones, in the long run the total amount of carbohydrate produced per unit of nutrient capital invested is greater than for deciduous trees.[14] But evergreen plants still face a problem in winter, and in order to compensate for the adverse effects of retaining leaves year-round, most have evolved a suite of water-conserving adaptations to carry them through times of drought.

Though the physiology of evergreen leaves clearly differs from that of deciduous ones, we cannot assume that evergreens remain photosynthetically active year-round merely because they remain green. In fact, winter photosynthesis seems to occur only under the mildest of conditions, such as in coastal areas. In most cases, evergreens enter a period of photosynthetic dormancy in the fall—apparently a prerequisite to timely freezing acclimation—and break this dormancy only after a period of several months. In the early spring months they are then usually able to produce and store limited amounts of starch reserves to support the later flush of new needles. This advantage is not unique to evergreens, though. Many deciduous trees, most notably the aspens, contain significant amounts of chlorophyll in their inner bark tissues and are likewise capable of limited photosynthesis, without leaves, under favorable circumstances. This may be true of the deciduous conifer larch as well, since it, too, contains considerable amounts of chlorophyll beneath its rough outer bark. While the significance of bark photosynthesis to the overall economy of the tree may be minimal, this illustrates well a point that will come up repeatedly in other discussions: There are many different solutions in nature to a common set of problems.

It is clear by now that leaves are not disposable, easily discarded accessories of the plant. The leaf sits at the end of a vascular system that extends without interruption through the twig to which it is attached, along successively larger branches through fork after fork, down the trunk, and out the main artery of a twisted and bent root to the smallest root hair probing the soil for moisture and nutrients. The leaf is one endpoint of a thread that resonates throughout the entire tree. As a molecule of water evaporates from the leaf, a tug is felt at the most distant root tip. The shedding of leaves, then, must be accomplished without disruption to the system—without exposing open-ended vascular tissue to uncontrolled water loss or entry of pathogenic bacteria and fungi. It is an operation tantamount to self-amputation, and it is performed with a surgical precision equal to the task.

Like all other preparations we have seen, leaf abscission, the actual breaking away of the leaf, occurs by design, involving many structural refinements at the cellular level. Most simple leaves like maple, birch, aspen, or cherry break away at the base of the petiole or leaf stem. Compound leaves like ash, sumac, or elderberry detach at the base of the leaf as a whole, but may also separate at each individual leaflet. In either case, the break occurs within a well-defined abscission zone, in which cells have been specially modified.

The abscission zone consists of two microscopically distinct regions of cells: a layer in which the actual separation takes place, and a protective layer that seals the exposed surface after detachment. The separation layer is generally two to several cells thick and is structurally weak. Unlike most mature cells throughout the stem, those in the separation layer generally lack lignin, a strengthening material normally deposited in the cell walls. Only certain of the vascular cells may be lignified, but then these are exceptionally short, which renders the tissue weaker than normal. The separation layer may be formed as early as the time of leaf development in the bud, or only shortly before abscission.[15]

Abscission usually begins with a number of chemical changes in cell walls, starting from the outside and proceding toward the center of the petiole. The whole process appears to be under the control of the senescence-promoting hormone ethylene (which also regulates the dropping of flowers and fruits), and in some species by another hormone, abcissic acid, that is instrumental in shutting down normal growth.[16] First the calcium pectate that cements cells together is converted enzymatically into water soluable pectin, and then the cellulose of the walls themselves begins to break down by hydrolosis, assuming a gelatinous consistency. Eventually only the vascular connections remain intact, but, weakened by their abnormally short stature, are prone to mechanical breakage in the wind. Prior to their separation, however, and minimally functional by now, the conducting cells often become plugged by balloon-like projections from neighboring cells, through connecting perforations in their common walls.[17]

Shielding the exposed surface after the leaf detaches involves the formation of a scar tissue, primarily through the deposition of a fatty substance known as suberin on the inside walls of a protective layer of cells (which may itself be several rows thick). This is often accompanied by the infusion of wound gum into the fabric of the cell walls and into the intercellular spaces of the exposed tissue. Finally, beneath the scar tissue, living cells form a new cork layer that becomes continuous with the outside bark.[18] The seal is now complete.

· · ·

Leaves are not all that turn in the autumn woods. Behind the flamboyant show of golds and reds, there is a quieter but no less important change occurring as virtually every resident bird and mammal of northern latitudes undergoes a molt. Many of the birds have already completed the process by fall, having begun shortly after brood rearing, but for others, as well as for the mammals, the shedding old feathers and fur while growing out new winter covers coincides generally with leaf fall—and for a few animals, this means a color change as well.

As with the shedding of leaves, day length is the principal factor controlling the timing of animal molt. Norwegian biologist Per Host first demonstrated this in 1938 with captive willow ptarmigan, in which he was able to manipulate plumage changes at will. By extending day length with artificial light in late autumn or early winter, Host was able to induce an unseasonable spring molt (the production of summer plumage); returning the birds to a natural light regime resulted in a reversal of their molt to an all-white winter plumage. While held under near-constant warm temperatures, Host's ptarmigan completed two molts over four months, with changing day length their only cue.[19] Since then, numerous studies have been conducted with mammals to demonstrate a similar day length control over molt. Experiments involving surgical removal of the light-sensitive pineal (p. 8) have shown conclusively that melatonin released by the endocrine gland acts directly to control prolactin levels in the hair follicle—and reduction of the hormone prolactin now appears directly responsible for the initiation of autumn molt in many different mammals.[20]

The only visible evidence of autumn molt in most birds and mammals is a certain shabbiness or aged appearance as old feathers and fur detach and fall out in motley fashion, gradually revealing the new undergrowth. As ragged as the animal may look in molt, though, there is nothing random about the process. Feathers and fur are replaced in a set sequence, usually with side-to-side symmetry. Patterns may differ between species, however, often depending on the ecological niche of the particular animal. Mink, for example, which remain brown all winter, start their autumn molt at the tail. By the beginning of September (at mid-latitudes), coarse new guard hairs can be seen among the old underfur near the tip, and within a month molting of the tail is complete. Over the next several weeks, replacement of the summer pelage progresses along the hips, sides, and back, until finally reaching the head. Three months from the start of the molt, the mink has a thick, glossy winter coat to protect it against deep winter cold and occasional immersion in icy water.[21] Other mammals, however, may molt in opposite sequence. Facing a different set of challenges for winter survival, the snowshoe hare, whose pelage changes to white, begins replacing brown hairs on the head and neck in a mottled pattern, with the cheeks and forehead remaining dark until fairly late. The feet, chest, and undersides then gradually whiten, leaving until last a broad streak of brown on the back that visually splits the image of the hare when viewed from behind,[22] rendering it difficult to see especially under patchy snow cover conditions.

Birds face a special challenge during molt in that replacement of flight feathers potentially hampers their mobility. Among insect-foraging birds, the larger wing primaries and tail feathers tend to fall out first, one or two at a time from each side, with regrowth quickly following in the same sequence so that flight is relatively unimpaired. Many water birds, on the other hand, lose all their primary wing feathers at once so that they are flightless for a period, leaving them quite vulnerable. In this case they often retreat, for the purpose of molting (a phenomenon sometimes referred to as "molt migration"), to areas that offer more resources than their breeding territories, where they can forage sufficiently and escape predators easily, without taking flight.[23]

In the fall the loon came, as usual, to moult and bathe in the pond, making the woods ring with his wild laughter . . .

As I was paddling along the north shore one very calm October afternoon . . . having looked in vain over the pond for a loon, suddenly one, sailing out from the shore toward the middle a few rods in front of me, set up his wild laugh and betrayed himself. I pursued with a paddle and he dived, but when he came up I was nearer than before. He dived again, but I miscalculated the direction he would take, and we were fifty rods apart when he came to the surface this time, for I had helped to widen the interval; and again he laughed long and loud, and with more reason than before. He maneuvered so cunningly that I could not get within half a dozen rods of him. Each time, when he came to the surface, turning his head this way and that, he coolly surveyed the water and the land, and apparently chose his course so that he might come up where there was the widest expanse of water and at the greatest distance from the boat. It was surprising how quickly he made up his mind and put his resolve into execution. He led me at once to the widest part of the pond, and could not be driven from it . . . It was a pretty game, played on the smooth surface of the pond, a man against a loon.

But after an hour he seemed as fresh as ever, dived as willingly, and swam yet farther than at first. It was surprising to see how serenely he sailed off with unruffled breast when he came to the surface, doing all the work with his webbed feet beneath. His usual note was this demoniac laughter, yet somewhat like that of a water fowl; but occasionally, when he had balked me most successfully and came up a long way off, he uttered a long-drawn unearthly howl . . . I concluded that he laughed in derision of my efforts, confident of his own resources.

—HENRY DAVID THOREAU, 1854[24]

Ground-dwelling ptarmigan, during autumn molt, often move from their summer territories to elevations above treeline, where snow patches may help conceal them as they change gradually to an all-white plumage. As the molt begins, body feather replacement is sporadic, with new white feathers dispersed among the dark, first at the head and neck and soon throughout the breast and back areas. This white mottling amid the ground-colored summer plumage effectively breaks up the ptarmigan's outline to potential predators (though in some locations, ptarmigan mortality may be high during the autumn hawk migration). As the season progresses, the new patchwork plumage gradually spreads to meet the expanding solid white of the ptarmigan's underparts. Once molt is completed, the bird usually moves back to its former territory, where shrub cover again becomes a primary consideration.[25]

Molting to white for winter would seem to have an obvious advantage for many animals, since cryptic coloration aids both predator and prey in their respective challenges. But a closer examination raises several interesting questions. Why, for example, does the arctic fox turn white when it has few predators to worry about and not much to sneak up on at this time of year? The arctic fox mostly scavenges in winter (often cleaning up after polar bears), or sniffs out voles and lemmings from under the snow cover, obviating the need for camouflage. If blending in with one's surroundings were particularly important to foxes, we might expect that the red fox—one of the most ubiquitous of northern mammals—would turn white for winter, too. And why, also, do the smaller mustelids—the least, short-tail, and long-tail weasels—that spend most of their time hunting the dark recesses of the snowpack in winter turn white, but not their close relatives the mink and marten that are more active above the snow? Is it possible that, for some animals at least, the primary advantage of molting to white is insulative? Lacking the granular pigment melanin, which imparts color to the darker summer plumage or pelage, the shafts of white feathers and hair are left hollow and therefore could be expected to serve as better insulators. As early as 1927 the Canadian biologist Seymour Hadwen noted "air bubbles" in the white hair shafts of snowshoe hares and concluded, against all popular belief, that "color changes are primarily for the summer exclusion and winter retention of heat"[26] (which may also help explain why arctic hares on Ellesmere Island in the Canadian Arctic remain conspicuously white all summer against a snowless background).

There seems little question that increased insulation is the primary function of the autumn molt, for all nonmigratory birds and mammals of

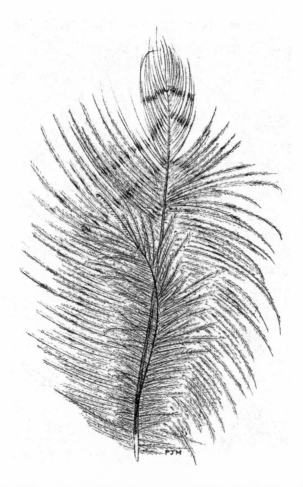

FIG. 1. A ruffed grouse "afterfeather" grows out of the inner shaft of a semiplumulaceous (downy) body feather. Produced during the autumn molt, these added feathers contribute significantly to the increased insulation of the bird's winter plumage.

northern and mid-latitudes are faced with a heat balance problem at this time of year (see p. 82), and regardless of whether or not a color change results, the outcome of autumn molt in every case is significantly greater protection from the cold. A bird may replace its summer plumage with 50% more feathers, decreasing its rate of heat loss proportionately by trapping more air close to its body. In many species this added plumage is produced largely through the growth of "afterfeathers"—auxiliary feathers that emerge from the inner shaft of a main contour (body) feather, sometimes even from down, often growing to half or more the length of the principal shaft (Fig. 1).[27] Similarly, the autumn molt in mammals can more than double the amount of underfur lying beneath the outer coat of guard hairs. In a manner analogous to a bird's producing afterfeathers, the secondary hair follicles that give rise to a mammal's underfur increase considerably the number of hairs produced in the fall from a common canal (as much as 48% more in our mink[28]), resulting in a significantly denser winter coat. The underfur is often crimped as well, adding to its air-trapping ability, and in many mammals the protective guard hairs increase in length to further improve insulation. For mid-sized and larger mammals, this added insulation in autumn can push the lower limit of their comfort zone (defined by the lowest temperature at which they can maintain body temperature without increasing metabolism) downward by 40°F or more.[29] Clearly, when it comes to preparation for winter, the autumn molt is as important to animals as the shedding of leaves is to deciduous trees and shrubs.

A Touch of Frost

&. Energetically speaking, there are two kinds of organisms in this world: those that can maintain stable (above-freezing) body temperatures by generating their own metabolic heat, and those that cannot. For the latter, which include the vast majority of species on earth, plants among them, frost is the preeminent danger of autumn in the north—not low temperature per se, but the actual conversion of water to ice. In its liquid state water is a major and essential constituent of all plant and animal bodies, but frozen, it can wreak havoc. The formation of ice crystals *within* living cells is inevitably fatal (the only exception presently known being within a certain specialized tissue of insects). The first rule of fall, then, is to avoid intracellular freezing.

The problem with freezing inside of cells appears to be a mechanical one; the growth of ice crystals disrupts membranes surrounding organelles, which leads to all manner of cytoplasmic chaos including uncontrolled enzymatic activity inside the cell. One of the first results observed with intracellular freezing in plants is the sudden leakage of ions like calcium and potassium from the cell, indicating that the outer plasma membrane itself has ruptured.[1] The evidence suggesting that it is actually ice crystal formation and not some other low-temperature effect that causes injury is clear; living cells can be immersed in liquid nitrogen (at $-320°F$) under carefully controlled conditions and then rewarmed slowly, without damage. The important difference between this procedure and normal freezing is that with rapid immersion in liquid nitrogen, water does not form ice crystals, but instead undergoes an apparently harmless amorphous solidification (sometimes descriptively referred to as "glass formation"). In one of the more exciting discoveries of plant physiology in recent years, a similar process was found to occur in nature under normal cooling rates. In balsam poplar, an extremely cold-tolerant tree species, glass formation in the intracellular fluid was observed at the relatively high temperature of minus $-18°F$.[2] But this remains a unique discovery. How often and in what other organisms this might occur is unknown.

One way to avoid the problems of intracellular ice formation in both plants and insects is to confine ice to the watery interstices outside the cells where problems of mechanical disruption are minimized. In the intercellular spaces of plants, ice crystals can grow to a size where they penetrate the very fabric of the cell wall, but a resilient plasma membrane lining the inside of the wall and a shrinking cell volume accommodate this growth without injury. This has an interesting secondary advantage. Ice, by its very presence, attracts water—a phenomenon having to do with energy gradients between the less active molecules on the surface of the ice and the more active molecules in liquid water. If an ice crystal forms first in an intercellular space, a likely possibility because the solute content of intercellular water is slightly lower than that inside the cell (hence its freezing point a fraction of a degree higher), then water will immediately begin to move from the adjacent cells toward the ice. As water passes through the semipermeable membrane surrounding the cell, it leaves dissolved substances behind in still higher concentration, lowering the freezing point of the intracellular fluid slightly more. By always keeping the intracellular freezing point a fraction of a degree below that of the water outside, plant cells are protected from internal ice formation without the production of specific antifreeze molecules.

The situation gets a bit more complicated with insects because the interstitial fluid is part of a circulatory system carrying oxygen, metabolites, and waste products throughout the insect body. The presence of extracellular ice crystals in this case disrupts normal transport, thereby interfering with chemical communication between cells. Additionally, when water freezes, solutes like sugars that are dissolved in it will be excluded from the ice crystal lattice and concentrated in the unfrozen water. An increase in extracellular solute concentration, while not a significant problem in the relatively pure interstitial water of plants, can create a condition of osmotic shock in insects. A certain amount of ice may be accommodated by freeze-tolerant insects, but others, though they may survive winter in very cold places, tolerate no ice whatsoever.[3]

The confinement of ice to intercellular spaces in plants or insects is not merely fortuitous circumstance, nor is the tolerance of ice in some insect bodies but not others. Long before the leaves begin to turn—indeed, well before the likelihood of first frost—processes were set in motion by which freeze-susceptible organisms began to acquire some degree of resistance. Much of the initial fall activity in plants and animals, then, occurs at a level we cannot see, involving the shutdown of certain metabolic functions and the startup of others.

. . .

It is well known that even the slightest touch of frost during the growing season will cause permanent injury to unprotected plant tissues. Yet the same frost that catches a plant unexpectedly at one time appears to have no effect at all a few weeks (or even days) later. Though there may be no difference in outward appearance, a bud nipped on a 30°F night in summer may withstand −30°F by early winter. What has changed?

The temperature at which water freezes in plant tissues varies little from one season to another; it is rarely more than a few tenths of a degree below that of pure water. Because water is confined within very small spaces in the plant, it can sometimes resist the phase transformation to ice until the temperature is several degrees below freezing—an effect known as "supercooling," where molecular attractions to surrounding surfaces temporarily prevent the rearrangement of water molecules into ice crystals—but the actual freezing point is still very close to 32°F.

What does change from one season to another is *where* the water freezes. We have already noted that ice formation within living cells is lethal, whereas ice formation outside of cells is generally not (unless it is carried to the extreme, but more on that shortly). So the difference between frost susceptibility in the summer and frost tolerance a few weeks later is usually only a matter of compartmentalization. In the one case, ice is forming both inside and outside the cell, and in the other it is occurring only outside. With water present throughout, how does this segregation of ice crystals occur? The key lies in membrane function, and this is where profound changes must take place within the plant prior to the first frost. This is where timekeeping becomes critical.

As some critical threshold of night length is crossed, an almost bewildering number of changes begin to occur in preparation for leaf senescence and winter dormancy. The onset of freezing acclimation, in particular, involves the manufacture and translocation of at least two messengers from the leaves that initiate changes throughout the plant. One of these appears to be a simple sugar whose role is obscure at best, while the other is the hormone abscissic acid, whose function is reasonably certain.

Evidence for the central role of a sugar molecule in freezing acclimation is circumstantial, but quite interesting: If a plant is held in complete darkness for a while, the onset of freezing acclimation will be delayed, even under cold treatment. But if sucrose is injected into it, acclimation will proceed as if the plant were experiencing normal, short-day light conditions. It is well known, too, that plants severely depleted in photo-synthetic reserves—sugars or their storage forms—will not acquire freezing tolerance. Just what role a sugar molecule might play in the in-duction process is not known at present, but sugars display cryoprotec-tive functions in other organisms that are undergoing freezing and may serve to alter membrane properties in some important way, perhaps by forming a protective shell around sensitive proteins.

The hormone abscissic acid, on the other hand, has two known func-tions that are directly implicated in the freezing acclimation process. First, abscissic acid is a universal growth inhibitor, synthesized in the leaf in increasing amounts as nights lengthen—and growth cessation is a prerequisite to freezing acclimation. In fact, anything that causes prema-ture growth cessation, such as a late summer drought, will promote ear-lier onset of freezing acclimation, but abscissic acid is the plant's control. Second, abscissic acid has a pronounced effect on the plasma membrane surrounding plant cells, significantly increasing its permeability to water. This is of paramount importance to the plant, for once freezing com-mences in the intercellular spaces of the plant (recall that the solute con-centration is slightly lower there, hence the freezing point a tenth of a degree or so lower), it is of vital importance that water be able to move freely across the cell membrane to join the growing intercellular ice crys-tal. As long as this can happen, the freezing point of the cell cytoplasm will stay safely lower than that of the intercellular water. If water is de-tained within the cell by too slow a movement across the membrane, only a slight additional lowering of temperature will cause it to freeze.

The initial short-day response of plants provides a narrow margin of safety to be sure, but it is normally sufficient (at the cooling rates usually experienced in nature) to see the plant safely through a light freeze. The first hard bite of frost then boosts the plant into a second phase of accli-mation, wherein a number of additional changes take place, rendering the plant hardy to the lowest temperatures it is likely to experience. A marked increase in lipid concentration observed in evergreen needles, for example, along with decreasing lipid saturation, suggests still other structural changes in the plant. Since membranes are constructed from lipids aligned in a double layer with proteins imbedded in them, the ob-served alterations of lipids might render the membrane more flexible under extreme low temperatures of midwinter.[4]

It is clear now that freezing acclimation is not merely the passive re-sult of growth cessation or exposure to gradually decreasing temperatures, but rather involves active metabolism just as do other facets of plant development. Yet requirements such as increasing membrane permeability are not compatible with the metabolic needs of the actively growing plant. So the two, freezing resistance and growth, are mutually exclusive. Like rival siblings they are occasionally at odds, and phytochrome is their mediator.

To the extent that insects share much the same environment with plants and are themselves small water-filled packages of living cells unable to generate heat, they face much the same problem as plants on the first cold night of autumn. Formation of ice crystals inside their cells will kill them. But the similarity between insects and plants ends there. Intercellular ice formation in insect bodies carries special problems, as we noted earlier, and must be carefully controlled—and this also requires considerable advance preparation.

Studies of insect freezing resistance have been complicated by the evolution of two entirely different approaches to the problem, and not without a significant amount of overlap and switching of strategies between species. Given the potential for intercellular ice crystals to disrupt normal metabolic function in insects—and some insects will remain active all winter under the snowpack in the constant presence of ice—it becomes necessary for insects to either avoid freezing altogether or to control the process closely. Both strategies, freeze avoidance and freeze tolerance, employ a complex set of biochemical, and sometimes behavioral, mechanisms, some of which are in response to decreasing daylength and others of which are stimulated directly by the low temperatures of autumn nights.

Freeze-avoiding insects go to the limit to prevent ice formation anywhere within their bodies. Theirs is not a strategy of avoiding cold exposure, for they are among the hardiest animals known (the arctic willow gall larva can survive body temperatures below −75°F), but they have no tolerance to the presence of ice internally and die immediately if freezing occurs. Freeze avoidance, then, requires a substantial freezing-point depression of the body fluids, but also involves a number of behavioral adaptations in order to prevent accidental inoculation of ice when the body temperature is very low.[5] The latter often comes first, and might begin with the selection of a dry overwintering place in order to avoid encountering external ice, which can induce ice formation in the insect body.

Because ice inoculation can occur in the presence of any small foreign body that might serve as a nucleus for an embryonic crystal, freeze-avoiding insects must evacuate their gut of all food particles, digestive microbes, and minerals or dust particles. Thus, they generally cease feeding as days grow shorter, with ample margin before the first frost threatens. Other potential ice nucleators that can not be evacuated, like lipoproteins, may then be chemically masked within organelles or membranes. In concert with this, the insect usually undergoes a programmed dehydration in which the total amount of freezable water in the body is reduced substantially. Not all of this water is lost entirely—some may be bound to cellular constituents, becoming an integral part of other macromolecules (hence unfreezable)—but the amount of extractable water may, nonetheless, drop from one-half to one-quarter of the fresh weight of the insect. Controlled dehydration also appears to be triggered by decreasing day length, but continues over a period of time long after exposure to subfreezing temperatures.[6]

Set up now for the onslaught of cold to follow, the freeze-avoiding insect begins to manufacture antifreeze molecules, including special proteins, that will eventually push the freezing point of its body fluid to the limit of its probable exposure. Stimulated by the first few nights in the 30s, the insect starts breaking down glycogen, converting it into alcohols like glycerol, sorbitol, mannitol, or ethylene glycol. Any or all of these are possible, although glycerol seems the favored one, and some insects become fairly well pickled with it. That champion of freeze avoidance, the arctic willow gall insect, may end up 20% by weight glycerol by midwinter![7]

Freeze avoidance is a strategy that clearly works for some, but it carries a risk. The only stable state in which water can exist at temperatures below 32°F is solid. Thus the freeze-avoiding insect exists in a metastable state most of the winter, where any disturbance can cause spontaneous nucleation of ice throughout its body. If something goes wrong in the deep cold of winter, flash freezing and instant death will be the result.

The surprising alternative to the freeze-avoidance strategy is not only to allow ice to form, but to help it happen and then control it. In what would seem an extraordinary temptation of fate, a great many insects in late summer prepare themselves for the coming frost by producing special proteins in their interstitial fluid whose sole purpose is to provide a nucleus around which ice can form more readily than would be the case without them. Ice-nucleating proteins, as they are called, apparently function in such a way as to disrupt the molecular forces that allow water in the interstices of the insect to naturally supercool (as in plants), rather than forming ice crystals as soon as it reaches the freezing point.[8]

The advantage of facilitating early intercellular ice formation is to avert the possibility of uncontrolled flash freezing and lethal intracellular ice formation—a constant threat whenever the insect is supercooled. To counter the problems of osmotic shock and disruption of circulation that we discussed earlier, the freeze-tolerant insect produces another antifreeze protein that acts to control the growth of intercellular ice by attaching to the small, induced crystals and blocking further addition of water to their surfaces. Freeze-tolerant insects then also hydrolyze glycogen to glycerol or other alcohols that serve to bind water, effectively reducing the amount of freezable water remaining and minimizing the dehydrating effects of ice formation.[9]

Perhaps the most remarkable aspect of all of this is the recent discovery that some insects, possibly even a great many of them, have the ability to switch from one overwintering strategy to the other,[10] each requiring, of course, entirely different preparations in the fall of the year. It is an extraordinary combination of measures in some very ordinary insects, coordinated splendidly by changing light and a touch of frost.

Reflections on the Pond

꿈 The pond also knows its season of change. As the days cool and nighttime temperatures nudge ever closer to the freezing mark, a process is initiated in lakes and ponds that will lead to an event of profound significance for the long winter ahead. It is a process related to the unusual physical properties of water, but it is one with far-reaching biological implications. To fully understand what goes on in the pond in autumn requires a brief look back in time.

When a lake or pond is heated during the summer, a vertical temperature gradient is established in which warmer water always overlies colder water. This is the consequence of a density gradient that parallels temperature—warmer water is less dense and floats to the top; colder water is heavier and sinks to the bottom. And though the actual density differences may appear minuscule—1 cc of water at 50°F weighs only .0003 g less than 40°F water—it is enough to resist the mixing of water of even slightly different temperature. Thus, when wind blows across a lake, the turbulence it creates will stir the water some, but its mixing effect will generally be limited to depths extending only a few feet below the surface (swimmers often notice the resulting abrupt transition to cooler, unmixed water beneath). The important biological implication of this is that once a lake or pond has become so stratified in summer, the mixing of oxygen, nutrients, and aquatic waste products also becomes restricted. Colder water on the bottom never mixes with the water above because its greater density prevents it from doing so. While oxygen and other gasses dissolved in the water will still diffuse in directions determined by their concentration differences, redistribution by this manner in water is a very slow process.

All this changes with a single important event in the fall. As surface waters begin to cool under the crisp night skies of autumn, the now denser water on top sinks. At first the effect of surface cooling is not very great. The heavier water can sink only to the depth at which it encounters water of the same temperature around it—a matter of inches, initially. In time, however, with continued cooling and subsidence from the surface, the upper stratum of water, now uniform in temperature, deepens. Eventually the effect of surface cooling will reach to the very bottom of the lake or pond. Once this happens, all density gradients will have disappeared and even a slight breeze setting water into motion at the surface will result in top-to-bottom mixing, a complete and even redistribution of resources before the lid of winter ice is clamped over the aquatic ecosystem. This is the process known as "fall overturn."

The temperature at which fall overturn commences and the length of time it continues will vary geographically and with depth of the water. Once the pond has reached the same temperature from top to bottom, the entire water body will cool slowly and continue mixing, until the water temperature reaches exactly 39°F. This is the temperature of maximum density for water, meaning that as water cools below 39°F it actually becomes lighter again. With a continued decrease in air temperature, then, colder surface water now remains on top, while the water at the pond bottom stays at 39°F (if this were not so—if colder water continued to sink—ice would form first on the bottom of the pond!). In time, the surface gradually approaches freezing, until, on some calm, clear night, the first transparent pane of ice forms over the pond. The lid is clamped on, and for many northern ponds it will remain in place for the next several months. The resources of the system, evenly distributed for the moment, are in finite supply for the duration of winter.

Buffered as it is now from the whims of winter weather, the pond under ice is not without its challenges, and some preparation in autumn is required of those animals that remain active. Aquatic organisms may not be handicapped by the high-energy demands that birds and mammals face in winter, for they have no set body temperature to maintain, but in cold water the chemical reaction rates that regulate metabolism shift into low gear, virtually immobilizing many. Generally speaking, for every 20°F drop in temperature, metabolic rates are cut in half—half for the change from, say, 75 to 55°F and half again for the drop from 55 to 35°F. Therefore, by the time ice forms on the pond, many aquatic organisms are already in slow motion, operating at only one fourth their normal pace. They are still responsive to physical stimuli, but are extremely lethargic and in a state of semihibernation.

With many organisms thus incapacitated, prey species would surely become easy targets for predators, were the predators themselves not also constrained by the cold water. Not surprisingly, though, several predatory fish (as well as some salamanders and crayfish) have evolved a means of overcoming this low-temperature restriction in order to maintain normal activity in winter. These species compensate for decreasing water temperatures in the fall through a general increase in the production of enzymes. More enzymes speed up chemical reaction rates, with the result that some fish end up able to sustain a level of metabolic activity in cold water that was previously possible only at considerably higher temperatures in the summer.[1] This gives the predatory fish a distinct advantage over their prey—but the imbalance is only temporary. With a higher level of activity under the ice comes a major tradeoff, resulting in little risk of predators overexploiting their newly vulnerable prey base.

The greatest problem that active animals face in the winter pond is a gradually diminishing supply of oxygen. The slow but continuing respiration of aquatic plants and other organisms throughout winter, particularly of bottom decomposers, results in the depletion of oxygen at the lowest depths, with the deficit extending over time higher and higher in the water. This eventually forces active predatory fish to move into upper strata, or toward inlets where stream water provides the only fresh input of oxygen into the system. Most of their prey species, however, the semihibernators, conserve oxygen through their inactivity and, as a result, are able to remain near the bottom where they tolerate oxygen deficiencies with little impairment. Most, in fact, have the ability to shift into anaerobic respiration when the oxygen is exhausted, deriving their maintenance energy by breaking down glycogen to lactic acid without need for oxygen. These organisms may eventually face the problem of excessive lactic acid accumulation in their tissues, but they have evolved ways of dealing with this, too, by converting the lactic acid to ethanol and either using the ethanol as an additional small source of energy or diffusing it into the water.[2] Thus, when the going gets tough, it is the lethargic species—the easy prey—that seem to have all the advantage.

A few aquatic organisms that inhabit bottom sediments around the perimeter of the pond or in stream banks may actually experience subfreezing temperatures. The lower surface of an ice cover, where it anchors to shore, represents a freezing plane that extends far enough into the sediments to affect organisms that might be buried there. While field evidence is limited, indications are that some invertebrates are able to migrate deeper into the soil, keeping ahead of the freezing plane as temperatures decrease through autumn, but inevitably others end up trapped there and perish for lack of evolutionary experience with subfreezing temperatures. A few aquatic insects, however, including mayflies, caddisflies, and stoneflies, are freeze tolerant, permitting some ice formation in their bodies but keeping it under control in the same manner as their terrestrial counterparts. (Deep supercooling as an over-wintering strategy [see p. 34] is not found in aquatic insects because of the constant threat of ice nucleation from their external environment.[3]) Generally, however, these insects avoid freezing conditions if possible.

Frogs and turtles that bridge aquatic and terrestrial environments may also end up exposed to subfreezing temperatures. Four North American frogs and at least one species of turtle in its juvenile stage overwinter on land, often with little more protection from deep winter cold than a scant covering of moss, leaves, or sand. And in the case of the land-hibernating frogs, their water-permeable skin and relatively large size virtually preclude their avoiding internal ice inoculation with the first hard frost of autumn. In spite of such certainty, though, these frogs show no advanced preparation for freezing, yet withstand it re-markably well—to a point.

As soon as ice begins forming in the frog's body, usually within the muscle tissues of its legs, the frog responds with a rapid conversion of liver glycogen to glucose (or glycerol in the case of the grey tree frog) and dumps it into the bloodstream in extraordinary quantities, some-times exceeding by 200 times its normal blood sugar level.[4] Concurrent with this is a pronounced increase in heart rate, stimulated by the release of latent heat with freezing (the conversion of water from liquid phase to solid liberates 80 calories of heat for every gram of water frozen), and resulting in the rapid distribution of glucose (or glycerol) throughout the body.[5] The high concentration of solutes in the blood does not act as an antifreeze, but rather appears to protect vital tissues from damage in the presence of intercellular ice crystals.

If freezing continues, both heart rate and breathing eventually slow to a stop. With upward of 60 to 65% of its body water frozen, the frog teeters on the edge of life, kept alive only by the imperceptible anaerobic metabolism of its energy stores. Remarkably, however, within an hour after thawing the heart resumes beating, and 6 hours later it is back to normal.[6]

The ability of these frogs to withstand freezing of such a large frac-tion of their total body water without sustaining tissue damage is, in it-self, truly extraordinary. But the fact that the frog shows no advance preparation for freezing is even more astounding, and invites question as to what adaptive benefit this seemingly risky strategy might confer. One answer, at least, may lie with the reproductive habits of the frog.

Because the frog is able to slip in and out of the freezing state without having to go first through a period of acclimatization, acquiring freezing resistance (usually involving loss of some other functional ability), and then deacclimatization, it can take advantage of marginal climatic circumstances when others cannot afford the risk. Becoming active early in the spring, as ice is just beginning to retreat from the shorelines of breeding ponds, these frogs start courting while nighttime temperatures still fall below freezing. This allows them to be more opportunistic and gives their offspring a head start, with lower predation and reduced competition for limited resources in ephemeral spring ponds. The frog's ability to withstand freezing is limited, however. So far as is known, no species is able to survive temperatures below about 18°F. This means that an autumn season with temperatures cold enough to penetrate the shallow depths at which most individuals hibernate, or a winter with little protective snow cover, can be fatal. Yet theirs is a proven strategy, for both the wood frog and the spring peeper overwinter in some of the coldest places in North America.

Only recently has it been discovered that painted turtle hatchlings also spend their first winter on land. Eggs deposited in sandy depressions above the waterline of a pond normally hatch in autumn, but the newborn turtles remain buried until the following spring, and like the land-hibernating frogs, they usually freeze with the first penetrating frost of the season. Unlike the frog, however, the turtle hatchling, in the first days of its life, synthesizes ice-nucleating proteins in its liver and distributes them to all extracellular fluids before blood flow ceases, seeding ice crystals, but keeping them small to prevent risk of supercooling and flash-freezing (recall discussion on p. 35). The turtle also produces glucose and other cryoprotectants to minimize later problems of cellular dehydration and membrane damage. As freezing occurs, then, ice forms first on the turtle's outer skin, and then grows inward toward the body core until eventually all circulation is cut off. Breathing and heartbeat cease, and only minimal brain wave activity persists until the newborn turtle thaws in the spring.[7]

After their first winter, painted turtles lose all ability to endure freezing and for the remainder of their lives overwinter on pond or river bottoms where oxygen deprivation, rather than ice formation, becomes their principal challenge. To conserve oxygen, turtles in hibernation enter a state of metabolic arrest in which energy use by cells is greatly reduced. But adult turtles in autumn also store large reserves of carbohydrate fuel, namely, glycogen, and this, as we have seen already, can be broken down in the absence of oxygen to provide energy through anaerobic respiration (p. 38). Adult turtles can endure extraordinarily long periods of time in this state, countering the potentially debilitating effects of lactic acid accumulation by buffering the acid with calcium and magnesium from their own bones or shell.[8] For the painted turtle, winter in the pond represents a remarkable turnabout from its first autumn spent frozen in the sand above water.

II.   JOURNEYS

Down the Long Wind

To a dweller in a city or its suburbs I heartily commend at this season the forming of this habit,—to look upward as often as possible on your walks. An instant
suffices to sweep the whole heavens with your eye, and if the distant circling
[hawks], moving in so stately a manner, yet so swiftly, and in their every movement personifying the essence of wild and glorious freedom,—if this sight does
not send a thrill through the onlooker, thn he may at once pull his hat lower over
his eyes and concern himself only with his immediate business. The joys of nature are not for such as he . . .   WILLIAM BEEBE, 1906[1]

&. Autumn in the arctic tundra is a bittersweet time. What the
growing season lacks in length, the continuous light of summer gives
back in part to produce an abundance of fruits, and autumn is the time
of reward. Dwarf heath shrubs turn color with reddening leaves and ripening propagules, and animals everywhere fatten on the bounty of
cranberries, crowberries, bilberries, and the like. But the autumn days
shorten with pressing rapidity—6 minutes, 7 minutes, 10 minutes—and
so there appears also a certain anxiety at this time of year. Activity increases; alertness is heightened; feeding is more frequent, almost continuous, almost gluttonous. So quickly the season passes. A sharp cold
front, a penetrating frost, and overnight the demeanor of everything
changes. For many, there is now an urgency to leave.

On the stony bench of an ancient shoreline a Northern wheatear
snaps at a few more berries and then abruptly stops, as if suddenly and
completely satiated, as if listening to a call no one else can hear. In an in-
stinctive second condensed from centuries of evolution, the bird takes
flight, and in its actions there is a deliberateness that speaks of great
purpose. As I watch it disappear, I know that I am witnessing the start
of a most remarkable phenomenon.

The wheatear flew straight west, undeviating, over the rolling gray
and russet tundra. It was soon joined by others. They continued west,

out over the dark and misty Seward Peninsula that bulged obtrusively into the Bering Sea. Straightening the undulating shoreline as they followed it, the small flock of birds headed for open water, leaving North America behind.

Soon over the Siberian mainland the flock veered slightly from its westward course and started a gradual, protracted sweep southward. Hour after hour in the darkness of the autumnal night they flew, stopping by day only to feed. Their flight carried them beyond the vast Siberian coniferous forest, past Lake Baikal, past the River Ob, the Kirghiz Steppe; across the parched Takla-Makan desert, the Arabian Peninsula, the Red Sea; onto the African continent, 7,000 miles from their birthplace. With an innate time and direction program—aided by sun compass, by stars, by the earth's magnetic field—this small group of birds from the treeless, rock-paved Arctic navigated over open water, high mountains, featureless deserts; across the Nile swamps, the dark equatorial forests, the great African savannahs; to a destination in southern Africa that they knew only from an ancient map, a map encoded in their genes, handed down through all their ancestry.

The wheatear family heirloom consisted of more than one map, however. Individuals that inhabited the eastern Canadian arctic set their course toward the rising moon instead, launching off the southern tip of Greenland for a trans-Atlantic crossing to England, thence to Africa. By either route, the wheatears of North America take the long way south (Fig. 2).

. . .

As migrants go, the wheatear's journey is not exceptional. There are several hundred bird species that make ordinary feats out of the extraordinary. Of some 9,000 species worldwide, roughly half engage in migratory flights. Even the "permanent" residents of many northern or mid-latitude locations shift their ranges seasonally through short migratory movements. The chickadees at a feeder in winter, for example, are not normally the same ones that frequent it in summer, but are usually birds from more northern areas that have moved a short distance southward (the summer residents also having shifted their range for winter). A great number of species are long-distance migrants, however, traversing a good portion of both the northern and southern hemispheres in their journeys.

Curiously, it is often those that breed in the remotest parts of the far north that cover the greatest distances—and not merely because they have the farthest to go to reach warm climes, for they often overshoot

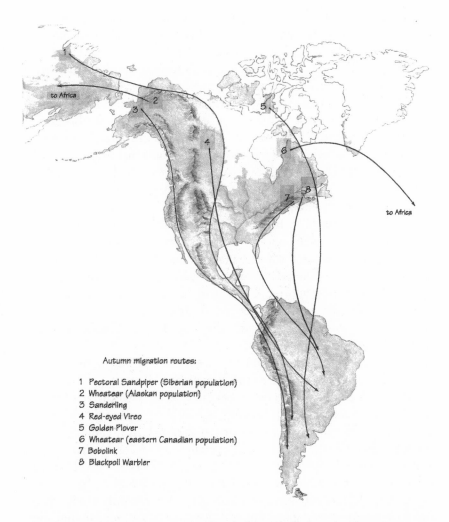

Autumn migration routes:

1 Pectoral Sandpiper (Siberian population)
2 Wheatear (Alaskan population)
3 Sanderling
4 Red-eyed Vireo
5 Golden Plover
6 Wheatear (eastern Canadian population)
7 Bobolink
8 Blackpoll Warbler

FIG. 2. Migratory pathways of several North American bird species illustrate common land and transoceanic routes to the southern hemisphere. Widely separated Alaskan and Canadian populations of the Northern wheatear, as well as Siberian populations of the pectoral sandpiper, are among the few that cross over between the New and Old Worlds, remaining faithful to the routes by which ancestral generations arrived at present breeding grounds.

the equatorial regions in favor of more southern parts of Africa or South America. The American golden plover, for example, funnels out of the arctic over the Canadian Maritime Provinces to navigate open water all the way to Venezuela and then on to central Argentina. Swainson's hawks, drifting out of northern British Columbia, ride the updrafts of the entire Rocky Mountain-Andean cordillera to winter also in Argentina.[2] And the record holder for distance, the Arctic tern that breeds in polar regions throughout the north, travels to the opposite end of the earth to summer again in Antarctic waters. Some terns, it is thought, even circle the Antarctic continent in their travels. Annual flights often span 25,000 miles over a 10-month period, and the terns do this every year, some of the true survivors for 25 years or more. Peter Berthold, a long-time student of bird migration, has done some interesting calculations on the lifetime distance covered by Arctic terns and estimates that individual totals may exceed three times the distance to the moon![3]

In many respects, though, it is the small land birds that are the most impressive migrants. The tiny ruby-throated hummingbird, weighing less than one-sixth of an ounce, regularly crosses the Gulf of Mexico in an 18-hour nonstop flight, requiring (according to Berthold) something on the order of 3.2 million wing beats![4] The blackpoll warbler, at half an ounce in weight (one eighth the size of an Arctic tern), flies 2,500 miles one way, much of it nonstop over the Atlantic between the coasts of New England and Venezuela, to winter in Brazil. The Swainson's thrush leaves its breeding grounds in the Canadian and northern New England coniferous forests to migrate all the way to southern Argentina. And then, of course, there's the Northern wheatear.

What makes the Northern wheatear's journey noteworthy, apart from the distance it covers, is its fidelity not only to ancestral wintering grounds in Africa, but to the ancient routes by which it connects with them. Sometime in the distant past, populations spreading northward out of Africa to breed apparently found their way to the New World via the northern British Isles and Iceland, while others, expanding their range throughout Asia, eventually crossed the Bering Strait into Alaska. The two North American populations thus established still retreat annually along routes by which their respective predecessors arrived. (The pectoral sandpiper of northern Arctic fringes has done the opposite: Today's Siberian populations, which had originally expanded their range across the Bering Strait from North America, migrate back to the Alaskan mainland to join others for their autumn journey to South America.[5])

Regular migratory crossings between the New World and the Old are relatively uncommon. Long transoceanic flights in particular, such as those made by wheatears, blackpoll warblers, and several other land birds are truly formidable, requiring extraordinary fuel supplies, flight efficiency, and aerobic endurance. For this reason, most migrants stay within continental boundaries, often following coastlines or major topographic features like north-south river valleys or mountain chains, perhaps as much for navigational purposes as for their physical advantages (as in the case of raptors following thermal currents along mountains). In North America, for example, major migratory pathways parallel the coasts, the Appalachian and Rocky Mountain systems, and the Mississippi drainage. Though movement often occurs as advancing waves along a broad front, large numbers of birds converge southward in autumn and eventually merge into eastern and western "superhighways" that then follow the islands of the West Indies and the Sierra Madre of Mexico into South America. A number of species, however, do cross the Gulf of Mexico directly for the Yucatan Peninsula or the coast of Venezuela (see Fig. 2). In Europe, a similar funneling of migrants occurs toward southern Spain, where the Strait of Gibralter offers the shortest water crossing to Africa.[6]

The greatest risks in crossing large water bodies stem largely from inadequate preparation and disorientation during poor weather, and occasionally large numbers of birds, especially first-year migrants, are lost under these conditions. Catastrophic deaths amounting to several thousands of birds have occurred during autumn migration over the Baltic Sea, coinciding with periods of particularly low visibility in dense fog lasting for several days. Even along the east coast of the United States, losses of young songbirds during migration have been estimated as high as 10%, due mostly to disoriented offshore movements. Waterfowl occasionally suffer large losses as well, even though they (presumably) can rest on the open seas. Juvenile mortality among barnacle geese crossing the North Atlantic between Spitsbergen and Scotland in some years approaches 35%.[7] But whether over sea or land, migration is an inherently dangerous occupation. Lighthouses, communication towers, and other tall structures often kill numbers of birds in nighttime collisions. Hunting pressure and heavy predation as migrants are funneled into narrow corridors may also claim large numbers. A Scandinavian researcher recently estimated that as many as 10% of the finches resting in southern Sweden fall prey to raptors during autumn migration.[8] Loss of habitat in wintering areas, inadequate search images for food in unfamiliar territory, and new diseases and parasites constitute still other hazards to migration.

Given the inherent risks, one might reasonably ask why migration has evolved as an obligate seasonal behavior in more than 4,000 bird species worldwide, involving an estimated 50 billion individuals per year. To anyone living in the north, the answer seems obvious: At a time when the majority of insects are dormant and hidden, when aquatic resources are out of reach under ice, when the thermal air currents needed by many raptors for efficient hunting are absent, numerous bird species are obliged to migrate. Even where substitute foods are available, most warblers, flycatchers, soaring hawks, shorebirds, and waterfowl have become too specialized in their feeding habits to alter their prey base and must move in the fall to areas that permit continued exploitation of familiar resources.

But the picture blurs a little when we look more closely at other migrants. Why, for example, does the rufous hummingbird leave southern Mexico in January to reach Alaska in mid April, and then time its return trip along the crest of the Rockies and Sierra Nevada to reach the Mexican border again by late July? And why does the common nighthawk gather in the skies over northeastern Arizona as early as mid August to begin its journey back to South America? Our food shortage hypothesis will not explain this migratory behavior, nor will it answer the question of why birds return in the spring if, for several months of the year, the resources of their wintering grounds are adequate to support them.

Turning the question around, I once asked a young Ecuadorian naturalist why she thought so many "tropical" birds migrated north in March and April, when the equatorial climate was so benevolent (our northern birds migrating south for winter were her tropical birds migrating north for the summer). Her answer was simple: "So they would have more room to breed." It was an interesting point: From her perspective, migration was necessary for reproductive success. It was as clean and unarguable an idea as our winter escape notion. The fact is that breeding activities are extremely demanding, and many birds have evolved territorial and nest site requirements as exacting as their feeding habits. And because birds do make the return trip north in spite of the fact that their maintenance requirements are met in the tropics, we must conclude that migration is not simply a search for more benign climatic conditions, but rather has developed to meet seasonally changing needs.

Whether the selective force that favored the evolution of such behavior was a seasonally induced scarcity of nutritional resources or lack of a suitable place to breed and rear young is largely academic. Individuals that long ago wandered out of the tropics to breed found more concentrated food resources in temperate or northern regions and ultimately experienced greater reproductive success than their cohorts that stayed behind. And those nomads that returned to the tropics after breeding

Every withered blade of grass and every dry weed as well as pine needle, reflects the light. The lately dark woods are open and light, the sun shines in upon the stems of trees which it has not shone on since spring . . . The atmosphere is less moist . . . light is universally dispersed. We are greatly indebted to these transition seasons or states of the atmosphere, which show us thus phenomena that belong not to the summer or the winter of any climate. The brilliancy of the autumn is wonderful, this flashing brilliancy, as if the atmosphere were phosphoric.   —Henry David Thoreau, 1851

From the edge of the wood the field slopes downwards to the longpond, now covered with a haze in the sunshine. The rushes fringing its edge are rusted and bent like old Roman swords, the reeds like the spears of ancient Britons, thrown with Arthur's sword, into the lake. By the pebbled shore the water is pure and clear and gloomy, the sunlight showing the moist brown velvet of the leaves upon its bed. Quietly feeding in the centre, a dozen moor-hens send ripples to the side, each wavelet bearing a shifting line of light over the leaves as it travels forward. Yonder the sallows* have loosened their slips of leaves and the sunshine throws up their ruddy and yellow wands—broken segments of a rainbow trembling by the marge.   —Henry Williamson, 1921*

* Willows, especially *Salix caprea* of the Old World.

Maple leaves on running pine. Unlike the deciduous trees, running pine will retain its chlorophyll all winter (p. 21).

Fall aspens glow as if lit from within. West Elk Wilderness, Gunnison National Forest, Colorado.

A time to disperse: dried seedheads wait for a lift (p. 96).

Home for winter, the insect larva in this gall has already prepared its spring exit and is now ready for the deep freeze (p. 33).

Fiery willows cannot avert autumn frosts. Streamside plants and aquatic insects will soon have to cope with ice in their tissues (p. 29).

Deceptive calm: Cold autumn nights set water in motion, eventually overturning the entire pond (p. 36).

By sun, stars, and magnetic field, these cranes have just navigated from the Arctic to New Mexico, possibly for their first time (p. 61).

Prior to migration many waterfowl retire to secluded waters to molt. There they may become flightless for a month or more (p. 25).

found winter survival easier and thus lived longer to pass on their genes. The resulting habit was clearly successful—enough to have become permanently imprinted in the genetic manual of survival for a great number of species. So we arrive circuitously at a view expressed far more simply and poetically by an early twentieth-century ornithologist from Aberdeen:

> In any case we must not think of North Temperate migrants flying south at the end of summer because of any definite prevision of the winter. They know no winter in their year, as the poet accurately observes, and have never known any, unless in the case of partial and incipient migrants. It is not by taking thought that these
>
> > ". . . wild birds change their season in the night
> > And wail their way from cloud to cloud
> > Down the long wind."[9]

. . .

There is no aspect of migratory behavior in birds that fails to elicit wonder. The unpretentious wheatear, ready to depart the northern tundra on a marathon journey, has acquired in the short season between brood rearing (or its own birth) and migration a readiness equal to that of a great athlete and explorer combined. From the physiological demands of long-distance flight to the challenges of navigating unfamiliar routes, migration is an extraordinary undertaking.

If there is a clear beginning to preparations for migration, it is the noticeable increase in feeding activity that attends the shortening days of autumn. Prompted by the hormones prolactin (the same hormone that caused molt in mammals) and corticosterone (a hormone active in carbohydrate metabolism), migratory birds enter a period of hyperphagia, characterized by a greatly increased appetite in which they display a remarkable ability to accumulate energy stores.[10] An abundance of high-carbohydrate berries at this time of year often contributes to their cause, but, interestingly, it is not carbohydrates that they store. Long-distance migrants must be able to extract maximum energy from the relatively small fuel load that they are able to carry, and ounce for ounce carbohydrates do not deliver the punch. Because carbohydrates cannot be stored without water—about 3 units of water by weight for each unit of available energy—they are not weight efficient. The advantage of eating fruits, however, is that their high content of sugar can be quickly converted by the liver into energy-dense lipids, mainly triglycerides, and distributed to any one of approximately 15 fat storage depots located in subcutaneous tissues or in the abdominal cavity of the bird. Not only are lipids stored without added water, but as fuel for flight they are easily oxidized, requiring fewer intermediate steps than in the burning of carbohydrates, and yield about twice the energy of carbohydrates per unit weight.[11]

At a time, then, when insect availability is waning, it is not unusual to see warblers, thrushes, cranes, wading birds, even ducks, feeding alongside seed eaters on the ripening fruits of autumn. The omniverous American robin increases its fruit intake from about 10% of its total diet in the spring to as much as 90% in the fall. Though not nutritionally complete, berries are generally plentiful at this time of year and are easily digested, providing a greater return on foraging investment than would result from an increased effort searching for insects. Thus, during hyperphagia it is often the frequency of eating rather than the size of meals that increases—and in many species the greater food intake is accommodated by an enlargement of the digestive tract. Even the house wren, which apparently does not switch to berries but increases its intake of insects, undergoes a 22% lengthening of the gut at this time of year, enabling it to retain and digest the added food for as long a time as it would with lower feeding rates.[12]

The result of all this feeding activity is, not surprisingly, a remarkable weight gain in long-distance migrants prior to their departure. While the normal fat load for nonmigratory songbirds is about 3 to 5% of their lean body weight (increasing slightly in winter), migrants of comparable size may add an amount up to 100% of their fat-free weight, effectively doubling their mass before takeoff. Some species of waterfowl are able to build fat stores at a rate approaching 10% of their lean body weight each day, reaching their maximum load capacity in less than 2 weeks. On the basis of body size, flight speed, and estimated energy consumption rates while in the air, calculated distance capabilities for birds in such condition range from 600 miles for the ruby-throated hummingbird to 6,000 miles for some of the larger waders like curlews and godwits.[13] The physiological requisites of hyperphagia persist throughout migration, however, so birds that land and feed en route are able to replenish fat stores quickly.

The nonstop distances covered by many migrants not only require a high level of fuel loading, but also demand a powerful flight engine and aerodynamic efficiency. The large pectoralis (breast) muscles of long-distance flyers may comprise up to 35% of their total body mass, and they function at extremely high oxidative capacity. With small-diameter fibers and a dense capillary network, these finely tuned muscles are equipped for rapid delivery of oxygen to respiratory sites where the lipid energy is released. Hypertrophy of flight muscles, the significant increase of muscle mass, is also observed in many birds prior to and during the migratory period. This is due in part to storage of fat in muscle tissue, but also involves an increase in fat-free weight, accompanied by increased aerobic capacity for energy production.[14]

Generalizations regarding physical characteristics and aerodynamic efficiencies of long-distance migrants are not as easy to define, partly because wing and body design almost always represents a compromise between the competing requirements of a particular way of life. Adaptations, for example, that confer energy-efficient transport during migration may be incompatible with adaptations for foraging in densely vegetated habitat on summer or winter grounds. Common characteristics among migrants are further obscured by the fact that birds of vastly different ecological niches, from hummingbirds to hawks, complete long flights annually. Recently, however, researchers from Austria and Germany conducted an extensive analysis of 32 morphological features in migratory birds of diverse groups, and found that one particular aspect of wing design was of almost universal importance: Within markedly different groups of birds, relatively longer and more pointed wings with well-developed outer wing segments are displayed by almost all long-distance migrants.[15] The Northern wheatear, for example, has a wing length 25% greater than that of its nonmigratory African relative the Somaliland wheatear, with the difference being almost entirely in length of primary flight feathers. Of two closely related European shrikes, the migratory lesser gray shrike has considerably longer primaries than the nonmigratory great grey shrike.[16] Even within races of the same species this difference seems to hold, as in the case of migratory and sedentary forms of the rufous-collared sparrow[17] in South America (Fig. 3). This is not to suggest, however, that total wing area of migrants is larger than that of their nonmigratory relatives, for just the opposite may be true. Likewise, there is no demonstrable reduction in wing load (ratio of body weight to wing area) among long-distance migrants of any particular group.[18] Efficiency in flight, it appears, is attained primarily through a streamlining, rather than enlarging, of the wing.

Some additional aerodynamic efficiency may be gained just prior to migration through the molting of worn flight feathers. During the year,

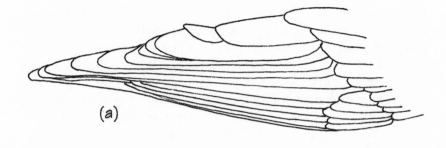

(a)

(b)

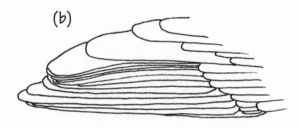

FIG. 3. Two races of the rufous-collared sparrow in South America illustrate a common feature among long-distance migrants. A highly migratory population from Chile (a) exhibits significantly longer primary flight feathers compared to a relatively sedentary population from Columbia (b). Such streamlining of wings, without increasing wing area, is typical among migratory birds of diverse types. [Redrawn from F. M. Chapman, "The Post-Glacial History of *Zonotrichia capensis*," *Bulletin of the American Museum of Natural History* 77(1940): 381–438.]

feathers suffer considerable abrasion against the ground or dense plant cover, and once damaged, can be repaired only through replacement. Most birds, therefore, molt completely at least once a year. Because replacement of old plumage requires a substantial investment of nutrient and energy reserves and is somewhat disabling, molt is normally postponed until late summer when the reproductive effort is completed, but food still relatively abundant. While this timing may be advantageous for other reasons, too (winter residents in northern areas add insulation against the coming cold through additions of feathers during molt; p. 27), for migrants the molt means sleek new wing feathers, perfect airfoils for the ultimate flight—and nearly all neotropical migrants complete their molt before departure.[19] It is the finishing touch, the trimming of the sails for the lonesome run "from cloud to cloud, down the long wind."

BY THE MERE

One morning in the hollows of the meadow land below the wood lay a silver mist. The sun sweeping upwards in its curve beat this away towards noon, but it was a sign. The fire of autumn was kindled: already the little notched leaves of the hawthorn were tinged with the rust of decay, already a bramble leaf was turning red: soon the flames would mount the mightier trees and fan their pale heat among the willows and ash round the lake, lick among the drooping elms and the lacquered oaks, and sweep in abandonment with yawning fire of color through the old beech forest.

. . . By the mere, about this time, hundreds of swallows would gather. Restlessly they clung to the sedges and the rushes, whose tops were beginning to brown and make faint whisper in the wind; now flinging themselves into the air, twittering, mounting high, wheeling and slipping, now descending like a shower of iron darts to the border of the lake . . .

As the days went on the swallows were more anxious. The cuckoos and the swifts had long since departed. Here and there in the fountain-shaped elms a yellow patch of leaves showed like a spilled plash of water imprisoning a sunbeam. The peggles on the hawthorns were reddening; and waving their pennants in the wind the dry rustling of the sedges came across the water with excited notes of the swallows. As one swept by, low in flight, the deep blue of his wings, their exquisite and soft uplift, the delicacy of the forked tail, were reflected on the surface; with a sighing sound he would pass and his liquid image glide on the mirrored surface below him.

The mornings were chilly, but the vagrants continued to hold their parliament in the rushes. Insect life was on the wane . . . Every passing of the wind beckoned a forlorn following of leaves from the trees; a spider seeking hibernation threw a prospecting line of silk against the face. With a tired sound the starred sycamore leaves . . . fluttered to earth; by listening closely it was possible to hear the stalks break from the twigs. Flittering like chafer beetles in a dusky summer night the vaned seeds risped and whirled away from the parent trees. As yet the conflagration had not caught the forest, only isolated flames browned a beech tree, scorched the branch of an ash with yellow or made buff haze in the distant oaks. It seemed as though the funeral pyre of dead summer would blaze in majesty only when the swallows had left.

One afternoon their shriller notes told that the hour was approaching. So eager were they among themselves that I was permitted to approach within a yard of them . . . Suddenly with a rush of wings they swept up, soon to become a smudge against the sky. But the wind was not favourable . . . for they returned to the sedges that never ceased to shiver of coming dreariness. The autumnal air was tranquil in its silence and solitude; the wings of gnats dancing their mazy columns assumed in the sunshine a fairy semblance. Over the waters sped the swallows, taking the last banquet, for once the long journey were commenced no halt would be made for food; the thousands of miles over sea and land must be passed without falter, urged and directed by the ancient instinct developed long before Zoroaster came from the plains of Iran with his Magian worship.

The next morning when I went to bid them farewell the lake was deserted . . . During the night a wind had risen, and they had fled before it to the warm south.

—HENRY WILLIAMSON, 1922[20]

What voice speaks to the bird that says "Go now—this is your time.
Fly quickly while the north wind blows and the new moon ebbs
to let the night stars show?"[21]

There is little question that the timing of fall departure is under a high degree of genetic control—that the biological calendars of migrants are ready-marked by generations of ancestral habit. This is strongly suggested by the nonsynchronous departures of birds like the rufous hummingbird and common nighthawk, in which neither weather nor food shortage appears to trigger the start of their southward migration (p. 50). But the most dramatic evidence of genetic control comes from wild birds kept in enclosures completely isolated from outside influence. Unaware of seasonal progression and unaware of the activities of their free cohorts, caged birds nonetheless exhibit all the characteristics of migratory preparation and departure, including wing fluttering—a display of stationary "flight" exhibited in captivity—while facing the compass direction that their normal migratory route would take them. That this behavior of captive birds is timed precisely with the activities of their free-ranging equals can only mean they are hearing the same ancestral call. But in the wild there is another voice speaking, too, for genetic instructions only open the window; ultimately, the individual must assess local conditions and decide when to fly. The actual start of migration—the moment of departure—may be determined by a number of variables like body condition, food supply, weather, or even phase of the moon.

Because many of the risks of migration, including premature energy depletion, are associated with adverse weather conditions, we might expect that weather is of major influence in the timing of departure. While birds clearly react to locally measurable variables like temperature, wind, or precipitation, they may, in fact, be responding to larger scale weather patterns that are unquestionably of great importance to their migratory success. For this reason, observers have long studied the relationships between numbers of birds departing or aloft and weather conditions, both on a local and synoptic (systemwide) scale, and decades of observation have yielded a number of generalizations about the atmospheric conditions under which birds will take to the air in migration.

Given the energetic advantages of flying with following (tail) winds, especially for migrants that regularly cross large expanses of inhospitable territory, it is almost axiomatic that peak fall migrations in the northern hemisphere most commonly occur on northerly winds, usually following the passage of eastward-moving low-pressure systems and their associated cold fronts. In a pattern typical of north temperate latitudes, low-pressure systems drifting across southern Canada and the United States regularly converge over New England and follow a path northeasterly, paralleling the coast of Canada's Maritime Provinces. On the leading side of the low, winds rise from the south, generally accompanied by falling barometric pressure and precipitation. With the passage of the system, northerly air flows ensue, accompanied by decreasing temperatures as the counterclockwise circulation of the low pulls colder air out of central Canada. This air flow is often reinforced by the clockwise circulation of trailing high pressure systems. Thus, in the wake of each passing low, nearly ideal conditions prevail for a southern departure, with both following winds and clearing weather occurring simultaneously (Fig. 4). Birds migrating in a southeasterly direction, then, tend to depart immediately following the passage of the cold front, when winds are predominantly from the northwest, while those flying south westerly will delay departure a short while until winds favor their travel.[22] Birds making the long ocean crossing from Greenland or Iceland to the British Isles often take advantage of the wind flow in

FIG. 4. By taking advantage of north winds after the passage of a cold front, long-distance migrants gain considerable energy savings. The numbers of birds and migratory insects flying south in autumn almost always rise following the passage of a storm, when both clearing weather and tail winds favor their movement.

polar air masses behind large Atlantic storms to gain maximum benefit from tail winds. Swans departing Iceland with the passage of a pronounced cold front have been known to reach Ireland, a distance of over 700 miles, in just 7 hours.[23]

This pattern is not completely consistent, however. While energetic considerations suggest that some small songbirds could not complete long flights across oceans or deserts like the Sahara without the benefit of following winds, other species may fly unhesitatingly against the wind, provided it is not too strong. Some diurnal migrants like hawks, cranes, and storks may even favor side or opposing winds if updrafts are present.[24] Swifts and swallows during fall migration regularly fly into headwinds, but may do so to facilitate feeding en route.[25]

With wind direction a major consideration, the next most important factor determining whether or not birds will take flight appears to be visibility. In dense fog or heavy cloud cover, the number of navigational aids available to a migrant is limited (see next section), which may result in straying. Investigators studying the arrival and departure of migrants on the Farallon Islands off the coast of California found that the presence of cloud cover diminished or eliminated the effectiveness of celestial cues, and at the same time reduced the ability of birds to detect or follow the coast, increasing their likelihood of drifting out over the ocean. But here, too, different species respond in different ways. Cloud cover will completely ground some birds, while others fly above or below it with little apparent loss of orientation (it is not unusual for some long-distance migrants to fly at altitudes approaching 20,000 ft). Fog generally has the same effect, though there is evidence that some migrants, reluctant to take off in heavy cloud cover, will do so in fog as though they were able to distinguish between the two. Soaring birds seem to provide the only constant: They rarely migrate during periods of fog formation, as fog is normally associated with atmospheric stability and the absence of updrafts.[26]

Precipitation also discourages flight, but does not suppress migration entirely. Departures are sometimes delayed until after rain stops, but once aloft, many birds will persist in light rains (though nocturnal migrants are less apt to continue than diurnal migrants). Birds encountering storms en route may attempt to divert around them, but are occasionally blown widely off course, sometimes resulting in spectacular landfalls of exhausted and disoriented birds. In late autumn, snowstorms occasionally promote "forced" departures, but overall, birds seem reluctant to depart during inclement weather, and radar studies reveal much reduced numbers aloft during periods of rain and snow or reduced visibility.[27]

There is presently no evidence that cold per se triggers a response in birds at the start of migration. While departures during fall migration are statistically correlated with cold or falling temperatures, this connection is likely an indirect one, tied to the shift in wind direction with the passing front. Similarly, a correlation between increased departures and rising barometric pressure might also be an indirect effect related to the passage of a cold front, but limited studies under controlled conditions indicate some birds may react to pressure directly—even to relatively small changes.[28] To the extent that rising pressure often portends clearing weather on a regional scale, it could be of adaptive advantage for a migrant to key in on this variable.

Two other aspects of migratory flight are deserving of brief mention here. One is the overwhelming tendency for normally diurnal species to migrate at night. Several ideas as to why this occurs have been suggested, including reliance on celestial markers for navigation (see p. 63), but the weight of evidence seems to point again to the high energy costs of migration and the advantage of maximizing daytime opportunity to refuel. Where inhospitable water or land crossings prevent stopovers, for example, a whole day of feeding at either end may be required to replace or secure adequate energy stores, necessitating nighttime migration for time savings. Even if the entire day were not required for feeding, the limited distance a bird might cover in the remaining daylight hours would greatly slow the progress of long-distance migrants. But nighttime flight, in itself, offers additional energetic advantages. The cooler, denser air at night favors lift, requiring less effort in wing flapping. At the same time, both vertical and horizontal air currents are reduced, resulting in less wind resistance and fewer course corrections. Furthermore, lower overall energy expenditure, coupled with reduced evaporation (a result of cooler temperatures and higher humidity at night), favors the thermoregulatory and water-balance needs of the migrant.[29]

Energy saving is also the most likely reason behind the flocking behavior of many migrating birds, as the air turbulence created by wing flapping favors those following the leaders. In the flight formations of geese, for example, the inboard wing of each bird is (ideally) aligned with an updraft eddy created by the individual in front of it, giving the trailing bird a certain amount of free lift. Evidence of this following advantage is often seen as birds in the lead frequently drop out to rejoin the flock near the rear, apparently because of tiredness. The actual energy savings incurred by this process depends on the number of birds in the flock (more is better) and the distance between them, with some overlap in alignment of wing tips and a slight stagger in flight level conferring the greatest power reduction. While theoretical power savings in the "ideal" formation can approach 50%, the unpredictable moves of the bird ahead and constant shifting to stay in the trailing eddy reduce benefits somewhat. Analyses of hundreds of formations suggest that the average savings for an individual is on the order of 20%—still a significant gain on a long flight.[30] Birds apparently can feel this energy savings and constantly adjust their flight attitude to maximize benefits.

While weather conditions significantly affect the efficiency and safety of long-distance flight, and appear to have an overriding influence on the numbers of birds departing on their southward journey at any given time, the effects of physiological condition are also important to the individual. A bird's readiness to continue its flight—the degree to which it has built up or depleted its fat reserves, the immediate availability of food and cover, and the dictates of its internal (genetic) schedule—will all play a part in whether or not a bird travels on a given date.

As the whirling winds of winter's edge strip the trees bare of their last leaves, the leaden sky of the eleventh month seems to push its cold face closer to earth. Who can tell when the northern sparrows first arrive? A whirl of brown leaves scatters in front of us; some fall back to earth; others rise and perch in the thick briars.

WILLIAM BEEBE, 1906[31]

. . .

If the entire scope of environmental awareness displayed by the migratory bird were not impressive enough, the sensory abilities required in navigating unfamiliar routes or featureless expanses of ocean, often at nighttime, stretches our imaginations to the limit. Consider the following, for example.

In one of the most remarkable accounts of homing ability ever reported, a Manx shearwater taken from a breeding colony in Wales and flown to Boston, where it was released, was recaptured at its nest site less than 2 weeks later. It had negotiated 3,200 miles of open water starting from an area well outside the species's natural range, had flown for days on end without landmarks in a direction perpendicular to its normal north-south movements, and had traveled at a speed that suggested unhesitating homing by the most direct route possible.[32] In a similar relocation experiment, golden-crowned and white-crowned sparrows that were flown from their wintering grounds on the coast of California to Baton Rouge, Louisiana, one year and to Laurel, Maryland, the next navigated over 4,000 miles of unfamiliar terrain to reach their usual breeding territory in northwestern Canada and Alaska and then returned to California, where they were recaptured the following winter.[33]

Studies like these demonstrate a strong goal orientation in adult birds, a drive toward specific geographical areas irrespective of starting point, that is somehow accomplished outside of their innate migratory program and without experience. When the Belgium researcher A. D. Perdeck intercepted 11,000 starlings on their way from Baltic coastal areas to wintering grounds in northwestern France and the British Isles, and detoured them by plane to Switzerland, the birds showed a remarkable response. Upon release, adults turned almost immediately northeastward and flew straight to their usual winter territory. First-year migrants, on the other hand, continued southwestward on their innate bearing as though they had never been interrupted, to end up in a totally new wintering area in southern France and the Iberian Peninsula. The following spring, however, the first-year migrants flew directly to their natal breeding grounds, rather than following a northeasterly bearing that would have taken them once again into unfamiliar territory. In subsequent years they returned to their newly adopted wintering grounds, now displaying specific goal orientation in their navigation (and remarkable short-term evolution in migratory habit).[34]

Evidence indicates that birds are instilled at birth with genetically encoded time and direction instructions by which they are guided initially to species-specific wintering areas. Experience may then modify the individual migratory route in subsequent journeys, but clearly the individual must posses some form of innate compass, in addition to having an internal clock capable of monitoring flight duration. At least three possible navigational systems, in addition to visual, have been identified so far, in which either the sun (or polarized light), the stars, or the earth's magnetic field may be utilized for direction finding.

Use of the sun for orientation was first demonstrated in birds by the German ornithologist Gustav Kramer in 1951. He placed European starlings in a circular enclosure having six windows to which he could affix mirrors for the purpose of altering the direction of incoming sunlight. When captive birds, exposed to normal light, exhibited migratory restlessness, fluttering their wings while facing in the direction of their usual migration (p. 56), Kramer abruptly changed the source of incoming sunlight by 90°. The birds adjusted their orientation to a new compass bearing at right angles to the first, and continued fluttering. When the source of sunlight was then reversed by 180° the birds turned completely around.[35]

Maintaining a constant orientation with respect to the sun's changing position is not as simple a task as it might appear, especially considering the wide latitudinal expanse covered by many migrants. When the equator is crossed, not only does the noon sun appear in the north instead of south, but the direction of its movement, as you face it, reverses. For wide-ranging birds, then, a solar compass not only has to be time (and sometimes direction) compensated for movement of the sun across the sky, but also must be constantly updated to track seasonal changes in the sun's path and the compass bearing of sunrise and sunset. Birds are apparently quite capable of time compensation through their internal clock, but adjustment to seasonal changes may lag behind the sun by a few weeks. Hence, accuracy of the sun compass in the domestic pigeon, for example, is thought to be to be only about plus or minus 10°, though that is usually enough to bring the bird into familiar territory, where visual cues might then guide it home.

An ability to orient by the sun has been demonstrated convincingly in about 10 species of birds, though most likely others use a solar compass in conjunction with additional cues. Solar navigation would seem to have its limitations in heavy cloud cover, although it has been demonstrated that many insects also dependent on solar orientation—honey bees, for example—can use polarized sky light (the alignment of scattered light in a plane perpendicular to the direction of solar rays) as a surrogate for the sun if as little as 10° of blue sky is visible.[36] Birds may have the same ability, though the evidence is tentative at present. In one study, migratory warblers exposed to artificially polarized light oriented themselves at a constant angle to the axis of polarization, the same angle that they maintained relative to the sun under natural light polarization. The solar compass is obviously of limited use to nocturnal migrants, too, though there is evidence that either the position of sunset glow or the pattern of polarized light at sundown can be used to establish initial direction or to calibrate some other compass.[37]

Given the propensity of so many birds to migrate during the night, it has long been suspected that stars provide a visual aid for navigation. Much circumstantial evidence supports the idea, beginning with the observations of another German ornithologist, Franz Sauer, who kept different species of European songbirds outside in cages exposed to the sky and noted their behavior at night. As long as stars were visible, their wing fluttering was oriented in the same direction as their normal fall migration, but as soon as the stars were obscured by cloud cover, they hopped randomly about their cage, or ceased activity altogether. Later experiments in a planetarium demonstrated that by shifting the north-south axis of the night sky 180°, the migratory orientation of Indigo buntings could also be reversed (just as Kramer had done with the angle of sunlight). It is now believed that birds use the axis of stellar rotation—the North Star or Polaris in the constellation Ursa Major—to determine polar direction, just as we do. If a bird reared without exposure to the stars is placed in a planetarium, and the planetarium sky is then rotated about some star other than Polaris, the bird will slowly learn to orient according to the artificially imposed stellar axis. An ability to recognize star patterns is apparently not important, but the bird must be able to view the sky area within about 35° of the pole star in order to discern the axis of rotation.[38] Evidence indicating orientation by stars has now been identified in about 20 species of bird (and a moth!), though the fact that many more nocturnal migrants show a decided preference for departing on dark, moonless nights suggests that the use of a star compass for navigation may be quite common.

In both practical and theoretical terms, the earth's magnetic field may provide the most useful means of all for referencing direction in unfamiliar territory: Magnetism cannot be obscured by cloud cover or darkness, and it does not require time compensation or correction for season. In addition to its north-south polarity, which, of course, has been employed in direction finding by human navigators for centuries, the earth's magnetic field has two other characteristics that may be of even greater importance to animal navigation. These are the vertical component of the field—the dip or downward angle of magnetic force with respect to the earth's surface—and the intensity or strength of the field, both of which vary between the equator and the poles.[39]

While in theory the properties of the earth's magnetic field are ideal for direction finding almost anywhere, the relative weakness of the magnetic force and the lack of any known magnetically sensitive receptors in animals discouraged biologists for some time from pursuing this line of investigation. By the early 1970s, however, it was becoming harder to ignore evidence that some birds could maintain an accurate compass bearing in the complete absence of visual clues. Furthermore, William Keeton at Cornell University had just demonstrated that the attachment of tiny magnets to the backs of homing pigeons completely disrupted their orientation. Birds that had small nonmagnetic brass bars cemented to their backs repeatedly homed in on their cages from some distance, while those wearing magnets scattered randomly.[40] Then in 1975, a marine biologist, who had observed magnetic responses in a bacterium he was working with, discovered a chain of tiny magnetite particles in the single-celled organism, all lined up perfectly with the axis of the earth's magnetic field. The particles, in effect, acted like a magnet with north and south poles, and aided the mobile bacterium in homing back to the bottom mud whenever it was swept up by water currents. In the northern hemisphere where it was found, the lines of the earth's magnetic force pointed both northward and downward, so that the bacterium, with the south pole of its magnetite chain at its forward end, would always find its way back to the mud simply by swimming along these lines of force.[41] This discovery sparked interest in other organisms of reported magnetic sensitivity, and soon magnetite was found in motile algae, salmon, dolphins, and other species, including honey bees— and homing pigeons!

For all the excitement it generated, however, the discovery of magnetite in birds soon proved somewhat of an enigma. Its presence was not consistent, and when both migratory and nonmigratory bird species were tested for remnant magnetism, no differences were found. This prompted a search for other possible magnetic receptors, which led to the curious discovery that magnetic direction finding in some species is light dependent and requires an intact retina. In bobolinks, another transequatorial migrant, both the optic system and the pineal gland become responsive to alterations in magnetic field when exposed to light, but lose all sensitivity in complete darkness.[42] With this discovery, the visual pigment rhodopsin soon became a prime suspect of magnetic sensitivity, though it is now beginning to appear that there may be two or more different magnetic detection systems in birds. In addition to magnetite and rhodopsin, melanin and certain biological radicals are currently being considered as candidates for magnetic sensing.[43]

Some form of magnetic compass has been found in all bird species so far examined, suggesting widespread use. At the same time, it appears that magnetic direction finding is inseparably linked to other orientation systems, with migratory birds relying on multisensory inputs in a complex and redundant way. The problem remains, too, as to what kind of atlas and geographic positioning system birds use to determine where they are at any given time, for a compass without a map is of little use. Though it may be a long while yet before the exact mechanisms of navigation are worked out, one thing has become clear: Birds know where they are on the map and know how to get where they want to go.

November 8, 1857. A warm, cloudy, rain-threatening morning. About 10 a.m., a long flock of geese are going over from northeast to southwest, or parallel with the general direction of the coast, and great mountain ranges. The sonorous, quavering sounds of the geese are the voice of the cloudy air, a sound that comes from directly between us and the sky, an aerial sound, and yet so distinct, heavy and sonorous: a clanking chain drawn through the heavy air. I saw through my window some children looking up, and pointing their tiny bows into the heavens, and I knew at once that the geese were in the air. It is always an exciting event. The children, instinctively aware of its importance, rushed into the house to tell their parents. These travelers are revealed to you by the upward-turned gaze of men. And though these undulating lines are melting into the southwestern sky, the sound comes clear and distinct to you as the clank of a chain in a neighboring stithy. So they migrate, not flitting from hedge to hedge, but from latitude to latitude, from State to State, steering boldly out into the ocean of the air. It is remarkable how these large objects, so plain when your vision is rightly directed, may be lost in the sky if you look away for a moment,—as hard to hit as a star with a telescope.

It is a sort of encouraging or soothing sound to assuage their painful fears when they go over a town, as a man moans to deaden a physical pain. The direction of their flight each spring and autumn reminds us inlanders how the coast trends.

—HENRY DAVID THOREAU[44]

The Improbable Flight
of Insects

&. For a long time we have been aware of the northward migrations of insects in spring. Every year from subtropical areas of Mexico a number of agricultural pests (the best known of the migrants) invade U.S. and Canadian crop lands. Similarly, from northern Africa mass insect flights regularly cross the Mediterranean into Europe. The scope of these invasions is such that they can be readily detected by radar. Until recently, however, these movements were thought to be one-way dispersal flights. Autumn migrations, if indeed they existed, escaped notice probably because movements were less synchronized than in spring, making it difficult to distinguish emigrants from summer populations in any one area. It was generally assumed that the progeny of spring nomads, having no history of cold exposure, eventually perished in northern winters; hence the springtime immigrants became known as "Pied Piper" migrants.

Though the "Pied Piper" notion of one-way dispersal remains popular (and may hold for a number of insects), there has long existed an example of true, two-way migration-and-return in insects. The convergent ladybeetle—that ubiquitous symbol of good fortune—has been known since the early 1900s to overwinter in large aggregations on certain mountain tops in the West, moving into lowlands during the spring and returning to their mountain retreats again in late summer or fall. While such flights are not particularly long (though 100 miles for a small insect is no short hop), nor always north-south in trend, they are seasonal, directional, and purposeful, which is the essence of migration as we have used the term. And they appear to be driven by food shortage in much the same way that the disappearance of specific prey forces many birds to migrate south for winter.

Ladybeetles are nearly obligate aphid predators, and they are born with voracious appetites. The larvae hatch from eggs in March or April, and for the better part of their first month do nothing but consume all the aphids they can find. They then pupate briefly, only to emerge one week later as hungry adults with the same single-minded purpose. If the adult continues to find sufficient food it will to grow into a reproductive state, but if aphids become scarce because of heavy depredation by the earlier larval stage, the adult, rather than starve, will migrate back to the mountains. There it will build up fat reserves by feeding on pollen, and then enter diapause—a state of arrested development—to overwinter inactively in the company of thousands of others. The release of an attractant pheromone by the beetles aids in the maintenance of large aggregations, which may facilitate reproduction in the spring when individuals might otherwise be too scattered to find a mate (clusters of ladybeetles also emit a musty odor that may serve to advertise their distaste, dissuading would-be predators).[1]

For some time the discovery of large overwintering populations of ladybeetles in the Sierra Nevada of California monopolized much attention. Aggregations had been described from which 600 gallons of the beetles could easily be collected (to be employed elsewhere for natural pest control), each gallon containing approximately 70,000 beetles. But equally large landfalls of the convergent ladybeetle in autumn were familiar to local residents along the shores of the Great Lakes and nearby bodies of water. From August through October, live beetles would wash upon the beach in such numbers that 5,000 to 10,000 individuals could be collected in an hour's time. These were clearly in a migratory or overwintering condition, lacking reproductive competence and having significant fat reserves. Empty guts also indicated that they were not feeding while in transit (recall discussion on p. 33 regarding gut evacuation and freezing resistance). The shoreline was not their overwintering site, however. The beetles had apparently fallen over the water and floated to shore, where they regrouped. Within 2 to 3 weeks they would resume their flight to unknown destinations.[2]

While ladybeetles rely on winds to carry them to overwintering sites, the circumstances of their migration suggest a more than passive ride aloft. The strongly directional nature of their autumn flight requires some awareness of compass bearings, for the beetle must be able to sense wind direction high overhead in order not to be carried far from its desired destination. At departure, the insect rises nearly vertically 2,000 to 7,000 ft., well above its normal flight altitude, until it hits the 55°F ceiling, which apparently is the minimum temperature needed to sustain its flight. It then levels off and is carried in the wind stream to specific overwintering sites.[3] For the California beetles, winds must be westerly, but other populations differ in their requirements. Those in the plains of Colorado, for example, await easterly winds to carry them into the Front Range of the Rockies.

Similar tactics have been observed in the brown planthopper of mid-latitude China, which only recently has been shown to make decisive equatorward migrations in the fall. To initiate migratory flight, this insect climbs at a rate of about 40 ft per minute to a level of 1,500 to 3,000 ft, to catch northeasterly winds on the leading side of anticyclones centered over Mongolia or north-central China. These winds take the insect on a trajectory straight toward traditional wintering grounds in southwestern China. Under average conditions the brown planthopper can travel nearly 200 miles in a 12-hour flight. The timing of its departure is also under rather precise internal control. In mid September, when wind conditions are right, migratory flight begins at dusk, apparently determined by some critical light level. As autumn progresses, however, lower seasonal temperatures boost the illumination setpoint such that departures times gradually become earlier.[4]

Equatorward migrations in the fall have also been documented now for the potato leafhopper and two species of lepidopterans: the true armyworm moth and the black cutworm moth. All three display reproductive diapause typical of other fall migrants, and all three show strong directional flight toward the south in spite of numerous opportunities for northward movements on the leading side of approaching low-pressure systems. Like migratory birds, these insects consistently fly on northerly winds associated with advancing cold fronts (see Fig. 4), though their southward progress is interrupted as soon as colder air temperatures overtake them. Thus their fall migration is characterized by a series of short-duration flights on successive fronts.[5]

Impressive though these small insects may be, the truly remarkable migrations among their kind are carried out by certain of the butterflies. The painted lady, for example, by its flights of extraordinary distance has become the world's most cosmopolitan butterfly. Unable to either hibernate or estivate (lie dormant in summer), this insect is an eternal nomad, found almost everywhere except in South America and a few other tropical areas. Every spring in North America the painted lady migrates en masse out of Mexico to repopulate almost the entire continent—and every fall it beats a retreat, far less spectacularly, to southern areas. In Europe, flights of this species have been traced from Morocco all the way to Ireland—half the trip nonstop over water—and rarely on to Iceland for a total distance of 2,100 miles from North Africa. Mass migrations of the painted lady heading south in late summer have also been observed in the Old World.[6]

The painted lady is not alone in its wanderings. Every year the European red admiral butterfly spreads out of the Mediterranean region to breed as far north as the Arctic Circle—but in no stage of its life cycle can it withstand northern winters, so every fall new progeny return south with extraordinary directional persistence. Researchers in northern Italy, studying the autumn migration of this species over a period of many years, have remarked on the "extremely precise—close to straight line" route taken by the butterfly, "[its] global directional preference tending to remain relatively constant day after day and in successive years."[7]

For persistence, navigational skill, and endurance, the North American monarch arguably reigns king. In a field of champions, the monarch stands out not only for length of migratory path, but for pinpoint accuracy in homing on a particular overwintering site. Every fall millions of monarch butterflies from northeastern North America undertake a journey to a small target 2,500 miles away in the volcanic mountains west of Mexico City—and not a single one of them, nor any of their immediate predecessors from the preceding generation, has ever done it before. Yet the trip is accomplished with unfailing fidelity every year. It is a most improbable feat.

In truth, no single monarch from northern areas ever completes the entire round trip. Their annual migration is instead something of an intergenerational effort, an accomplishment of the species rather than of the individual, and this is what makes it all the more remarkable. When adult monarchs mate and then leave their overwintering roost in the mountain firs of central Mexico, they begin what amounts to a migratory relay for the United States. Laying eggs along the way as they encounter the milkweed plant (to which their life cycles are obligately tied), adults die, larvae hatch, grow, pupate, and new adults emerge to continue dispersing northward. Each succeeding generation pushes the species front farther. Finally, the northernmost populations reach their summer range where one or more additional generations may then be produced . . . and here's the remarkable thing: Not a single butterfly from a summer generation shows any genetic propensity to migrate. Yet, one more preautumn generation, and the migratory urge is reinstilled. The last born has all the instructions to begin an incredible journey southward, without experience, without a relay. The last born will itself complete the entire return in autumn—or die trying.

The determination of a migrating butterfly is unmatched. Once it sets its bearings on a course, it almost refuses to deviate. Unlike the insects we have discussed so far, butterflies will struggle directly into a head wind for days, flying only yards above the ground. And stories abound of their efforts to overcome, rather than detour around, obstacles in their path. Torben Larsen, one of the most experienced of butterfly migration observers, recounts watching butterflies ascending past the ninth floor window of his Nairobi, Kenya, hotel room to clear the roof of the building, rather than making a simple pass around it. Butterflies crossing the Mediterranean, on encountering a ship, will flutter straight up one side, across the deck, and down the other.[8]

Large geographic obstacles are no deterrent either. Puzzled once by the passage of some 4 million white caper butterflies in an unusually narrow front (only 12 miles wide!), Larsen got in his car and backtracked the migratory column to investigate. From the map it appeared that they were being funneled between the Aberdare Range and Mt. Kenya, but everything in Larsen's experience told him the butterflies should be flying *over* the mountains—not squeezing through lower passes. When he reached a 10,000-ft plateau of the Aberdare Range he found his answer. The migrating caper whites that were flying over the range were encountering on the leeward side a marked drop in temperature, below the critical threshold at which they could maintain flight, and were raining out of the sky, stopped by a physiological barrier rather than a physical one. No funneling was occurring at all, but only those butterflies that came through at the lower elevation were able to keep going. The others would simply have to wait for warmer temperatures.

The stimulus for migration, the imprinting of directions—especially for a goal-oriented species like the monarch—and the navigational tools used by butterflies are still unknown. Flying mostly during daylight hours, there is a good possibility that either the sun or polarized sky light is used in conjunction with magnetic direction finding to keep the insect on course. Any way you look at it, the flight of butterflies is most improbable, especially considering, as the noted expert Robert Michael Pyle once put it, that the computer in charge consists only of a minuscule number of neurons that passes for a brain in the insect!

Walking the Whole Way South

❧ Migration is generally affordable if an animal can fly. For most mammals, however, long-distance travel gets expensive—and the smaller the mammal, the more prohibitive the cost. The amount of energy expended in moving a given distance overland is about five times greater than that required of flying, for animals of comparable mass, and on the ground caloric expense escalates quickly as the animal's size diminishes. A red squirrel, for example, might spend 10 times more energy per unit body weight than a full-grown deer in covering the same distance.[1] Without the ability to accumulate high-density fat stores rapidly, as do birds, most mammals might find the cost of migration prohibitive—which is probably why only a handful of species regularly move between widely separated summer and winter territories.

The exceptions are few. Bats generally migrate—some, like the silver-haired bat that summers as far north as the Arctic tree line, covering impressive distances annually (see p. 89)—but they do it, of course, by flying. Elk in the western United States migrate from summer territories in the mountains to wintering pastures in the valleys, but seldom do they range more than 40 or 50 miles between the two. Bears may also travel annually to favored fall foraging locations well outside their individual territories, but these, too, are relatively short movements. (In one part of the southern Rockies, bears at the end of August regularly migrate in a direct and purposeful fashion, over 3 to 5 days, to a particular oak-brush area where they spend most of September and October before finally retreating to their preselected denning sites—and this regardless of food availability within their own territory.[2]) This leaves caribou, then, as the outstanding migrator among northern land mammals.

The barren-ground caribou of North America are gregarious animals that spend winters within the shelter of the Alaskan and Canadian boreal forest, but move to very specific locations on the Arctic tundra for calving. The distance between wintering and calving grounds may be on the order of several hundred miles and the caribou's rate of travel such that the better part of the year—sometimes 10 months—may be spent moving to and fro. Their reasons for such nomadism can only be understood in the light of their life history.

Caribou are specialist ruminants that subsist for a large part of the year on a diet primarily of lichens (the so-called "reindeer moss"). Lichens are abundant throughout the north, but they are slow growing and prone to rapid depletion if browsed too heavily in one location. The ingrained nomadism of caribou thus seems to have evolved partly as a means of avoiding overgrazing in large herds. Roaming widely within expansive territories also assures a continuous food supply in winter, and for this purpose the caribou are uniquely equipped with broad hooves that are ideal for both walking on and pawing through snow. But this does not by itself explain the directional nature of their annual movements. Nor does escape from frigid temperatures explain it any better, for caribou are extremely cold tolerant and, given shelter from wind provided by the forest, endure the deep freeze of boreal winters rather easily.[3] So we are once again left with a question, the same one we asked of birds wintering in the tropics: If the boreal forest provides all the needs of caribou in winter, why do they leave it in summer?

The answer has a lot to do with insects. The scourge of life in the northern summer is biting insects—billions of them. Large herds of caribou are plagued not only by blood-sucking mosquitoes and black flies (mosquitoes can kill an unprotected newborn calf within hours), but also by botflies and warble flies that deposit eggs on any exposed skin, and whose larvae then tunnel under the skin to feed and grow on the animal's own fat reserves. The only relief available, it seems, is to stay close together and move out onto the coastal tundra where exposure to wind helps keep insect numbers down, at least part of the time. Even at that, the energy expenditure incurred in fighting insects is extreme—greater, in fact, than that which is required to maintain body heat in the cold of winter—and by the end of the insect season on the tundra, animals are exhausted and often calorically bankrupt. This, in turn, explains much about the nature of fall migration in caribou.

The beginnings of fall migration are vague and uncertain, somewhat like the warming up of a symphony orchestra, where the only prelude to the great music that will eventually follow is the random noise of various instruments. The music is at first disorganized and incoherent, but imagine that without a clear beginning, without the usual pause and signal from a conductor, the discord of unsynchronized notes gradually folds into complete harmony and we suddenly become aware that a concert is underway. This, in essence, is how the great caribou migrations of fall come together.

By early August on the tundra, the torment of insect harassment gradually lessens and the caribou herds begin to relax their agitated behavior. Organization loosens, stretches, starts to come apart at the edges as individuals wander—tentatively at first, then more deliberately. The advantages of close company under the onslaught of insects slowly give way to the need for unrestricted grazing—the need to rest and recoup energy reserves severely depleted. The daily cost of fighting off insects during the summer—of constantly moving, often at the expense of eating—has nearly doubled the caloric demands of winter.[4] So the caribou now begin to disperse with single-minded purpose, undaunted by the dangers of predators, uninterested in the company of others.

Their dispersion has an underlying direction to it, however. It is not completely random, but like a lazy smoke takes on a definite drift, in this case to the south. And when the dispersing caribou come up against barriers like unfrozen lakes or wide river crossings, they bunch up again—maybe several thousand strong, but only temporarily. Once across or around the obstacle they fan out again, almost shunning each other, their spreading front expanding in width at about the same rate that it drifts southward. Temporary aggregations may form at insect pockets, but for the better part of a month, the outstandingly gregarious caribou go their own way.[5]

By early September the first of the wandering caribou may have reached the treeline, and there they begin to bunch up again, drifting along the forest edge one way and then another, slowly gathering in numbers. Some animals will move into the forest, perhaps even penetrating a hundred miles or more, but only to turn again and rejoin the growing herd at treeline.[6] This is a leisurely time, if ever such exists in the life of a caribou: a time to put reserves back into the energy bank, for the true fall migration—and the mating season—has not yet begun. The orchestra has only been warming up.

In October, things begin to take on a more serious air. Though there has been some evidence of sexual activity in the gathering of the herd, the rut—the serious competition for mating opportunity—is only now coming into its own season (see following chapter). At the same time, a growing restlessness to move on to the winter range pervades the herd. It is never certain which will commence first. An early snowstorm might stimulate migration prior to the start of the rut—but in either case, once the rut is underway, the urge to migrate seems to strengthen,

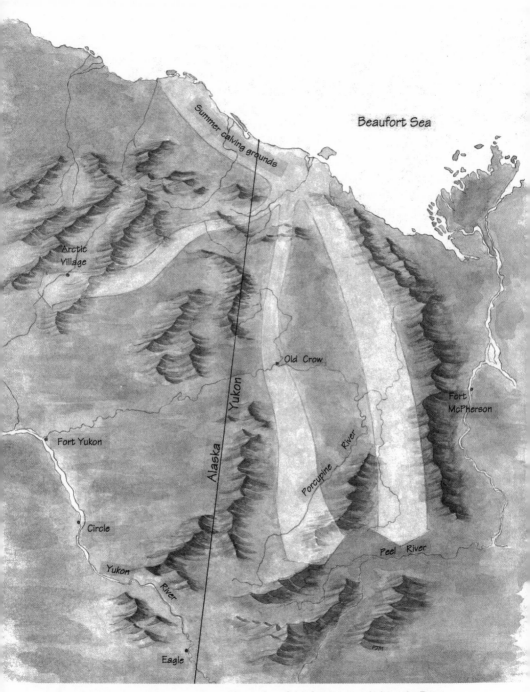

FIG. 5. After summer calving on the windy tundra of the Arctic coastal plain, the Porcupine caribou herd begins its southward trek of several hundred miles to overwinter in favored areas within the boreal forest. Traveling for nearly 5 months, caribou migrations represent the longest seasonal movement of any land mammal.

and movement deep into the forest becomes strongly directional and rapid. Across an advancing front that may span many miles, the caribou file into long columns, lines of concentrated movement separated by areas in which only stragglers may be found. By early December the caribou are generally in their wintering areas, one more trek of several hundred miles credited to their experience.[7]

Though individuals are not necessarily faithful to the same route throughout their lifetimes, certain travel corridors are favored, generally those that follow paths of least topographical resistance, often paralleling larger drainages (Fig. 5). And some major routes, judging from the early records of trappers and explorers, have been in use periodically since the early 1700s. Hudson Stuck, an early missionary to the Kutchin territory of interior Alaska, recorded near the end of the nineteenth century that the banks of the Charley River for a distance of 50 miles were "trodden hard and solid by innumerable hoofs of caribou," these of the Fortymile herd.[8] Even with the difficult times of late—with the interruption of traditional migratory pathways by advancing human development—the autumn movement of caribou remains one of the greatest shows in the world, for the migrations of this animal are as insistent as the species itself.

Deep Rut

The deer's horns have grown harder, and with the urge of autumn, they scrape
them on the bushes and small trees ("horning the brush" it is called) tearing off
the dried skin and sharpening the points . . . Like the deer, the elk are polishing
and sharpening their full grown horns to the destruction of lodgepole saplings
whose stripped, naked shafts tell us the elk are coming out of their seclusion. Finally on a cool, frosty morning in a retired mountain park a deep throaty roar
rings through the air, and is repeated again a half hour later. It is the clear, bell-
like challenge of the male elk, proud of giant strength.

MILTON PHILO SKINNER, 1924[1]

The shiftlessness of the caribou herd as they gathered at treeline—the seemingly aimless wandering of cows, the flare of hormones
and unprompted sparring of bulls, the fickle changing of allegiances
between small bands—belied a pattern and a seriousness of intent that
might have escaped notice altogether by those unfamiliar with the ways
of ungulates. To the caribou this behavior was orderly and purposeful, a
fundamental process in their annual cycle, played out by set rules. It was
the autumn mating period, the rut, and its timing was critical, as the
narrow window of opportunity for reproductive success was dictated by
the crucial spring calving time and a long gestation period. This is true
of all the large ungulate herbivores, whose calves must be born as early
in the year as the new growth of forage allows, in order that they have
sufficient time to develop fully before their first winter. This means that
mating must take place during a relatively brief period in the fall of the
year, or not at all.

The first hint of the coming rut is a heightened anxiety among
males that seems to attend the shedding of the velvety skin covering
the year's new antlers. Males begin in August thrashing brush and
small trees with their antlers almost as if annoyed by the strips of skin
they are trying to remove. They become intolerant of other males, and

their thrashing often takes on appearances of fighting, as if testing their strength against the indifferent trees before engaging another male in a serious contest for a female. As the season continues, the thrashing may become something of an erotic stimulus.[2] An additional behavior, the frequent habit among all deer of suffusing themselves with their own urine during rut, may be a form of scent marking, as if the male itself were the center of a moving territory.[3]

That all deer, including caribou, moose, and elk, should display a number of common rutting behaviors is related to the fact that all are polygamous and all males are faced with the same basic challenge of attracting a female and wining the right to breed with her. Polygamy generally arises out of either an imbalanced sex ratio, with more females than males in the population, or situations in which females congregate for reasons other than mating and males then attempt to monopolize them. Among the gregarious caribou and elk, these two conditions often act in concert. The sex ratio within larger herds of barren-ground caribou in Canada and Alaska is typically one male to every three females.[4] A similar ratio is found among elk throughout the Rocky Mountain region,[5] a circumstance that seems to arise out of a higher rate of mortality among bulls.

In many polygamous mating systems, males attract females by establishing and defending (against other males) territories with desirable resources into which females wander. Where control of high-quality environments is not possible, however, either because resources are abundant everywhere or, alternatively, are so scattered that their defense is not practical, a different strategy may be employed in which the male associates with a female (or group) and follows it wherever it wanders, defending the female instead of a specific territory. In this case, a female becomes, in effect, a moving territory to be guarded against competing males until she is successfully mated by her suitor.

Though female defense, especially when it entails herding a group of cows, appears on the surface to be under the control of the male, just the opposite is the case. Polygamy among deer is actually based on female choice, rather than on male dominance—a situation that has to do with greatly different reproductive stakes involved for the two genders. While a male can afford to be indiscriminate in choice of mate, the female cannot. If a male mates at all—not a certainty, because it must first gain access to females by competing with other males—it usually will mate more than once by virtue of the dominance it won through superior fitness, and thus it has a smaller stake in each mating because only one success is all that is needed to pass its genes along. Being indiscriminate in choice of female is not necessarily disadvantageous as long as at least one mating produces a fit progeny. Females, on the other hand, can almost always get fertilized because there is usually a willing male around, even if only a yearling of unproven fitness. But the stakes are much higher in this case because a single mating normally provides sufficient sperm to fertilize all of the female's eggs. Thus, it behooves the female to be discriminating in selection of a male in order to assure the most fit offspring. From the perspective of the female, the wrong choice of mate could prove disastrous.

In female defense polygamy, then, "defense" on the part of the male is nothing more than earning the opportunity to mate by showing superiority in warding off all potential competitors. In reality, the male is defending only itself and its breeding privileges. It is in the female's best interest, on the other hand, to foster as much competition as possible in order that the males sort themselves out, with only the fittest remaining dominant by the time the female is receptive. In a breeding system like this, the male is left with only two possible strategies, each very different from the other. The male may either advertise its own fitness in order to attract several females, as in the case of elk, thereafter maintaining and defending a harem, or it may seek out, stay with, and defend one female at a time until it has successfully bred. For the male there is little choice in the matter, however, as both habitat character and female behavior dictate which strategy will work and which will not.

The autumn rut in elk provides the classic example of how environment shapes behavior. In the relatively open forests of the western United States, females congregate in small groups to protect their young, often sacrificing food quality for the advantages of increased predator vigilance by many individuals.[6] Under such circumstances, harem formation and tending becomes feasible because bulls can usually watch all members of their band at once. But this also makes defense of females physically taxing because in open forests dominant bulls are obvious to other challengers, and cows are often attracted to other harems. To be successful in this situation, bulls must not only be more effective in attracting and retaining females, but must then have the endurance to repeatedly defeat other males in combat, while also keeping members of their harem from wandering off to join others.

For bull elk, advertising fitness entails vocalizations or "bugling," a repertoire of brays, squeals, and bellows in which a deeper and louder call is indicative of larger size and higher dominance rank, the yen of all breeding females. But vocalizations also attract challengers for their dominance—and cows are listening to other bulls as well. If a female perceives another as superior, it may attempt to switch loyalty, even seeking matings with more than one bull to assure a successful pregnancy. In order for a bull to maintain its harem, then, it must bugle more frequently to counter the advertising of rivals, and must continually reinforce the positive experience of its cows by freeing them of constant harassment from younger, more eager bulls. And even within the harem there is a jostling for rank as more mature cows vie for dominance over younger, inexperienced individuals. This puts the proven fit in line for first breeding, assuring their pregnancy and allowing their calves to be born earlier in the spring.[7]

Where other members of the deer family occupy denser habitat at rutting time, or maintain smaller female groups, harem tending becomes energetically impractical. Even caribou congregating at tree line are too loosely knit as a herd and too dispersed among the sparse trees to warrant such strategy. Under these conditions, courtship and mating behaviors are altogether different from those of elk, requiring that males seek out and stay with potential mates, one at a time, until they have bred. In the case of caribou, rutting behavior is directed at identifying females in estrus by "testing" their receptiveness via short pursuit and then identifying with and tending a single individual, presumably one that is in heat. These bouts of "driving" by the male are often chaotic and frequently interrupted by sparring matches with nearby bulls, during which the chase may be taken up by other males. But once tending commences, the male follows the cow wherever it goes and becomes uninterested in the activities of others, showing antagonistic behavior only toward individuals that approach too closely.[8]

In contrast, the mule deer, whitetails, and moose of densely forested areas all tend to be relatively quiet during the rutting season. Unlike the bull elk that advertises its position, the male's interest in this situation is to remain as inconspicuous as possible while tending a female, even to the point of hiding with the female and treating her so as to create the least possible disturbance in order not to attract competitors. This poses an interesting dilemma for the female, however, and a different kind of challenge to the male, since it is in the best interests of the doe or cow to do just the opposite—to incite as much competition as possible among available males through widespread roaming, noisiness, and conspicuous estrus. Thus the male in this case must work at preventing a female from running or leaving a hiding place. (Mule deer bucks have apparently answered the challenge of exciting and holding the interest of females by imitating the vocalizations of juveniles, appealing to the doe's maternal instincts. The most common sound uttered by a mule deer buck during the rut is a soft, low version of a fawn's distress call.[9])

Courtship and mating in a polygamous system, whether it involves male advertising and harem tending or the secretive pursuit and defense of one female at a time, is extremely demanding of the males. Many forego feeding during the rut and, as a consequence, end up calorically bankrupt at the end of the season,[10] putting them in a precarious position should winter and deep snow arrive early (see next chapter). Success in this system, however, is the ultimate measure of strength and endurance—the truest test of the male's ability to obtain sufficient nutritional resources for growth, to withstand the rigors of winter, and to avoid predators. And only the fittest are allowed to pass their genes on to the next generation. It is a system that assures continuance within the population of only the best survival traits.

A Dangerous Chill

If the covers could be taken off the fields and woods at this season, how many interesting facts of natural history would be revealed!—the crickets, ants, bees, reptiles, animals, and, for aught I know, the spiders and flies asleep or getting ready to sleep in their winter dormitories; the fires of life banked up and burning just enough to keep the spark over till spring.   JOHN BURROUGHS, 1924[1]

 Winter can be especially hard on mammals in the north country. Caught in a cold trap from which they cannot escape, their primary challenge becomes one of trying to hold body heat against the relentless vacuum of deep space. Almost everything the mammal does in winter is directed toward that end, but whether it succeeds or not depends largely on preparations made in the fall.

The problem confronting warm-blooded animals in the cold is not complicated: In order to maintain constant body temperature in the face of declining air temperatures, heat input must offset heat losses plus or minus any change in storage. It's the familiar balance sheet problem with expenses rising faster than income. For the animal, however, the energy budget *must* balance. Deficit spending beyond the limited fat reserves of most mammals or birds results in assimilation of body protein for maintenance needs, with death following quickly if relief is not immediately forthcoming.[2] To some extent, increased energy costs with the coming cold can be offset through behavioral adjustments—increasing insulation by erecting fur to entrap more air (or by nest building in some cases), and reducing surface area of exposure by lying, curling, or huddling with others. Gains can be effected by absorption of sunlight through basking, or by increasing metabolic heat input. In the field, however, both the problems of and the solutions to energy deficits tend to be considerably more complex than on paper.

From the standpoint of the animal it is usually of greater advantage, when the weather turns cold, to operate first on the heat loss side of the

*The light reflected from bare twigs at this season . . . is not
unlike that from gossamer, and like that which will erelong
be reflected from the ice that will incrust them. So the
bleached herbage of the fields is like frost, and frost like snow,
and one prepares for the other.*

—Henry David Thoreau, 1858

Most of us are still related to our native fields as the navigator to undiscovered islands in the sea. We can any autumn discover a new fruit there which will surprise us by its beauty or sweetness. So long as I saw one or two kinds of berries in my walks whose names I did not know, the proportion of the unknown seemed indefinitely, if not infinitely, great . . . As some beautiful or palatable fruit is perhaps the noblest gift of nature to man, so is a fruit with which one has in some measure identified himself by cultivating or collecting it one of the most suitable presents to a friend. —Henry David Thoreau, 1860

Crystal threat: The greatest danger with freezing comes from the formation of ice crystals in the cellular environment (p. 29).

In defense of its harem, a bull elk (behind tree) risks caloric bankruptcy by the end of the autumn rut (p. 77).

Geometry of hope: Casting their fate to the wind, these seeds gamble against heavy odds. Few will land at a safe site (p. 96).

Common burdock has found the secret. Seeds may travel farther by clinging to animals than by any other method of dispersal (p. 101).

Fresh aspen cuttings prepare the beaver lodge for a winter of "impoverished comfort" beneath the ice (p. 113).

Begging for harvest, piñon pines have formed an inseparable partnership with seed-caching jays, each depending on the other (p. 105).

Few symbols better capture the sense of autumn bounty than the traditional pumpkin patch.

Just as late-flowering goldenrods push summer into autumn, so wild apples offer a vestige of autumn to winter browsers.

ledger, for the measures just noted carry minimal energetic costs. Once everything has been done physically to cut losses, then the animal, in the face of continued temperature drop, must step up its rate of metabolism, with the increased energy demands being met either from food stores (including fat reserves) or increased foraging effort. Birds, of course, face the same challenge in winter, but have the advantage of far greater foraging mobility than mammals. For the latter, it gets more and more difficult to work on the income side of the budget when snow deepens and food sources become hard to find. And migration, as we have seen, is not an answer for many mammals. Even the caribou, for their long walk south, do not have an easy time of it in winter.

For mammals that remain active in winter, two aspects of fall preparation become critically important, one affecting energy loss and the other energy gains. We have already noted the importance of autumn molt in adding to the insulative value of fur. For animals like the caribou, changes in guard hair length and underfur thickness can be so effective in reducing heat loss, especially in combination with other behavioral adjustments, that stepped-up heat production may not be required until nearly −50°F.[3] This is true to a lesser extent in other deer that also acquire thick insulation during the fall molt. Yet, even with this advantage, food often becomes scarce enough (or is low enough in quality) that maintaining basal metabolism can become a problem, especially late in winter. Frequently the energy obtained through increased foraging effort is not enough to offset the increased costs of moving around in the snow, in which case the animal is better off resting and fasting. So the second critical factor in preparation for winter, particularly for large mammals, is the addition of fat reserves. All the large ungulates subsidize much of their activity during cold weather with energy gained earlier, their overwintering success often depending as much on the amount of fat added in the fall (or remaining after the rut) as on their pelage changes.[4] For smaller mammals, however, there are practical limits to the addition of fur or fat before mobility is impaired, often mandating a different approach to overwintering, and different preparations in the fall.

One possibility for cutting losses on the expense side of the energy budget is by reducing body temperature. Because the rate of heat flow from warm to cold objects is dependent in large measure on the temperature difference between the two, trying to maintain normal body temperature in winter is, from a purely physical point of view, self-defeating. If the animal's internal thermostat were turned down instead, the temperature gradient across its insulative coat would be lessened, slowing heat loss all the more. Benefits would include reduced maintenance needs, lower energy requirements, and less time spent foraging. It's a simple idea, with only one serious drawback: The mammalian nervous system does not function well at subnormal temperatures (hence the dangers of hypothermia), so an animal in this state would have to become completely inactive or dormant, yet still retain the ability to generate heat spontaneously in order to recover.

Notwithstanding such limitation, many small mammals, as well as a few birds, have evolved this capability for short-term benefits—a strategy known as torpor. Lowering body temperature by several degrees can get an animal through a cold night with considerable energy savings. (Note that torpor differs significantly from hypothermia. In the latter, metabolic rate increases in an effort to maintain body temperature, until, under extreme cold stress, neither can be sustained. The rapid drop in heart rate, metabolism, and body temperature that follows cannot be reversed without heat input from external sources.) A few mammals (but no birds) have taken this approach further, achieving a deeper depression of metabolic rate over extended periods of time—long enough to carry them through an entire season of food shortage and low temperature. These, of course, are the true hibernators, and theirs is a much more serious game. While relatively brief and shallow torpor bouts can be entered into spontaneously, successful hibernation requires significant advanced preparation.

Digging deeper into hibernation takes us into a much less familiar realm of biology where our concept of "warm-blooded" versus "cold-blooded" quickly runs into limitations. Birds and mammals have long been referred to as warm-blooded in the belief—generally correct—that they maintain a stable (warm) body temperature at all times. But now we are confronted with a paradox—a hibernating mammal that for a considerable part of the time lets its body track ambient temperatures nearly to the freezing point (or occasionally slightly below). Yet we still consider these mammals warm-blooded because in our simplified classification scheme we have no other category into which to place them. It is primarily for this reason that the term "homeotherm" has been widely adopted as a substitute for "warm-blooded." A homeotherm is defined as an animal that is capable of maintaining body temperature independently of its surroundings by producing heat internally through its own metabolism (the opposite—an animal incapable of generating significant metabolic heat and whose body temperature fluctuates with that of its surroundings—is a "poikilotherm"). The term homeotherm reflects the fundamental ability of all birds and mammals to control their temperature, but gets us away from reference to a specific temperature or inference that the blood is always warm, thereby accommodating the hibernator. Yet, even with our new vocabulary, the hibernating mammal seems a contradiction of terms, able to sustain low body temperatures as a poikilotherm would, without suffering the effects of hypothermia that kill most homeotherms, and yet unable to remain continuously in that state through the winter because certain mammalian functions require high body temperature, at least occasionally.[5] So the hibernator plays both ends against the middle, remaining cold and seemingly lifeless for days on end and then firing the furnace back up to normal, only to lapse almost immediately into dormancy again.

The requirement of periodic rewarming throughout winter, a process referred to as arousal (though it does not necessarily involve activity), places a theoretical, if not practical, constraint on the feasibility of hibernation for some animals. The amount of energy needed to raise the temperature of a 10-pound marmot from near freezing to normal, an increase of more than 60°F, is not inconsiderable, and all of the necessary heat must be provided through the metabolism of stored reserves—either excess fat as in the case of the marmot, or cached food. Though smaller animals would require less energy to arouse, they would be expected to have proportionately smaller reserves—and the tiny pocket mice, some of which remain in hibernation for 10 months, may experience more than 100 arousal episodes in that period of time.[6] Two additional factors may tax energy stores further: Temperatures that threaten to go significantly below freezing may prompt more frequent arousal (or require a higher degree of thermoregulation) than might otherwise be necessary, and a late spring may force the hibernator to subsist for an extended period of time on limited reserves left after a long winter. Thus, the practicality of hibernation as a strategy for coping with seasonal food shortage and low temperature may be limited by the availability of "safe" hibernacula and the amount of reserve energy an animal can put up for winter.

With increasing latitude, the availability of den sites in which interior temperatures are not likely to drop more than a few degrees below freezing must necessarily decrease. It is perhaps for this reason that hibernation as an overwintering strategy is far more common at midlatitudes, with few true hibernators ranging north of 55° latitude. The arctic ground squirrel is one of only a half-dozen exceptions, hibernating regularly in permafrost regions within burrows where unfrozen soil is an ephemeral condition at best. In this case the dormant squirrel is exposed almost continuously to below-freezing conditions over an extended period of time, but the arctic ground squirrel is also the only known homeotherm able to withstand subfreezing tissue temperatures without sustaining permanent damage[7] (though some bats are suspected of having the same ability; see p. 89).

The food storage requirements of hibernation may impose even greater restrictions on animals, not only confining this overwintering strategy to certain ecological niches (no strict carnivore hibernates), but also setting an upper limit to the size of animal that can hibernate. Whether an animal caches food in the hibernaculum or develops a thick layer of fat depends on some balance between total energy needs and storage constraints, with small animals limited physically in the amount of fat they can accumulate, and larger animals unable to stockpile enough energy to carry them through the winter. The Eastern chipmunk is generally considered to be the maximum-sized animal that can go through extended hibernation on cached food, while the marmot is thought to be the largest animal that can store sufficient fat to hibernate.[8]

The exceptions again prove interesting. Richardson's and Columbian ground squirrels hoard food in addition to storing fat, but it is only the male that does so. The larder in this case appears to be a hedge against energy shortage in the spring, when the male emerges early to await the appearance of females.[9] And on the size limitation, both brown and black bears would seem to explode our generalization. There is still much discussion, however, as to whether or not bears are true hibernators, with differences of opinion centering largely on body temperature as a defining criterion. A bear denned up for winter seldom displays a temperature drop of more than 10 to 12 degrees below normal—a long way from the temperature depression that most hibernators experience. It can be argued, however, that an animal of such mass, insulated as well as it is, has a considerable amount of thermal inertia in its favor and would not be expected to cool significantly in its den, even though its metabolic rate in winter may be only half its normal rate.[10] By any other measure, bears behave much like true hibernators, accumulating large stores of fat in autumn prior to entering their winter dens, and once retired, generally remain dormant for 6 months or more without emerging.

· · ·

As in migration, the two prerequisites for entry into hibernation are the involution (atrophy) of reproductive organs and fattening, for even the food hoarders must have ample fat to fuel initial arousal. And not surprisingly, both appear to be under the control of melatonin, mediated by shortening daylength and acting through the pineal (recall discussion on p. 8). The role of melatonin in the repression of sex hormones and atrophy of reproductive organs in both males and females is demonstrated conclusively by surgical removal of the pineal, blocking the secretion of melatonin. But unlike the stimulus for migratory preparation, temperature appears to be an important secondary trigger for both prehibernation activity and entry into dormancy. A sharp drop in temperature can override the action of the endocrine gland and induce reproductive atrophy, even with the pineal removed.[11]

Actual entry into hibernation is a far more complex process than might be imagined from observing the animal. While a mammal may lapse into the hibernating state from sleep, it is clear that hibernation and sleep are not at all the same thing. Brain-wave patterns characteristic of an animal in the hibernating state are distinctly different from those associated with normal sleep. In fact, some biologists believe that during hibernation a significant sleep deficit is accumulated and that arousal is necessary in order to satisfy the sleep debt (in homeotherms sleep is possible only at normal body temperatures).[12] Nor does entry into the hibernating state involve a simple passive release of body temperature control. Rather, the animal appears to actively depress metabolism, with heart rate, blood pressure, and oxygen consumption all dropping before a decline in body temperature becomes evident.[13] This is accomplished through controlled hypoventilation in which carbon dioxide is accumulated, leading to a condition of respiratory acidosis in which the pH of arterial blood is lowered. The effect of acidosis is then to depress metabolism, which leads to a reduction in body temperature. Lowered body temperature in turn has a positive feedback effect on metabolism, depressing normal chemical reaction rates, with the cycle driving temperature farther down.[14]

The pattern of body temperature drop as animals lapse into a state of hibernation differs among species, and even among individuals with varying experience in hibernation. Some, like the arctic ground squirrel, go through a series of "test drops" involving sequential bouts of torpor in which body temperature is lowered by 15 to 20 degrees over a period of a few hours and then returned rather quickly to normal. This may be repeated daily for as long as a week before the animal finally lapses into a deep state, where body temperature may fall 80 degrees or more in 24 hours and remain there for several days before arousal.[15] In a modification of this pattern, the California ground squirrel may take each test drop to a lower level during alternate torpor bouts, stepping down the lower temperature limit by a few degrees each time, before halting the decline with a bout of shivering. Individuals have also been observed to enter the hibernating state through a series of step-down plateaus, lowering body temperature a few degrees at a time, but holding each drop for a short while before continuing the process.[16] Still other species like the woodchuck and thirteen-lined ground squirrel may go through a steady and relatively rapid decline over several hours until they reach a deep hibernating state, though sometimes in an inexperienced individual periodic bouts of resistance temporarily slow or reverse body temperature during this process. In woodchucks, the descent into hibernation is frequently interrupted several times during the first few attempts, but animals that have been through hibernation many times do not normally interrupt the decline. Golden hamsters rarely interrupt the descent into hibernation unless physically disturbed.[17]

The timing of hibernation can vary widely among different mammal species. Once an animal is prepared, entry can be rapid and induced by a sharp drop in temperature or sudden depletion of food supply. Recently, on a frosty, late-September morning in Colorado, I removed from a live trap a least chipmunk that gave all appearances of being dead. On a hunch I held it in the palm of my hand and in the sun for some time while discussing with my field biology class the intricacies of hibernation (gambling all the while that the chipmunk was merely torpid, not dead). In about a half hour the chipmunk showed faint signs of abdominal movement, and within the next minute or so it stirred, sat up, and abruptly leapt out of my hand. Clearly its time for hibernation was near, and the frosty night was all it took to induce a state of deep torpor. Had the animal been in a burrow, rather than in my hand, it might well have lapsed into complete hibernation. Similar instances have been reported where food shortage, rather than cold, induced deep torpor in a ready hibernator.

Even within the same environment, different species may show significantly different timing in their entry into hibernation, which sometimes reflects differences in food preference and its seasonal availability. Several of the North American ground squirrels, for example, feed on a variety of plant materials, including seeds, fruits, and fungi, but prefer grasses and graze heavily on them in the earlier part of summer when they are most nutritional. These mammals fatten early and often are in hibernation by late summer. Food cachers, on the other hand, tend to be more dependent on seeds and generally remain active later into the fall when this resource is more abundant.[18]

Ultimately, then, the timing of hibernation depends on preparation—and so, too, does its success. Juveniles of a species often enter hibernation later than adults because their first year growth interferes with fat accumulation, and for the same reason they often suffer considerably greater overwinter mortality. Juvenile fatality among hibernating Columbian ground squirrels may approach 70%, compared to a high of 40% for adults, and in the woodland jumping mouse may more than double that of adults (the highest rate of juvenile loss yet reported is 93% for a population of Belding ground squirrels).[19] So it is a precarious business. Hibernation is a journey of a different sort, into a realm that at times is neither that of the homeotherm nor yet the poikilotherm. It is limbo, and it doesn't always work.

. . .

If any single animal exemplifies all of the processes and adaptations that we have been discussing in this section, it would be one of our northern bats, for here is a mammal that seems to have borrowed the best from all that nature has evolved. It flies, it migrates, and it hibernates. It is gregarious, polygamous, mates in the fall and births in the spring. And it is successful, for bats number 925 species worldwide—which means that one out of every five species of mammal on this planet is a bat.

While bats of northern latitudes share the same energy balance problems of all other homeotherms in winter (p. 82), they are, in spite of their small size, rather tolerant of cool temperatures. Big brown bats, for example, will remain active even after nights have dropped below freezing, and have been seen flying during snow flurries late in the year. And though data are scant, some species of bats have been reported to withstand subfreezing body temperatures without harm.[20] But northern bats have another problem. They are specialist feeders—aerial insectivores like the flycatchers—and the disappearance of flying insects must ultimately force their migration to winter quarters (though at least one species, the gray bat, begins migrating in August).

Members of seven different genera of bats migrate moderate to long distances, with several other species moving shorter distances. The record holder appears to be the hoary bat, which summers as far north as the upper end of Hudson Bay and occasionally in parts of Alaska, and winters in the tropics or subtropics. The European noctule commonly migrates 600 to 900 miles, but two outstanding records for this species include autumn flights out of southern Russia to Istanbul, a distance of 1,100 miles, and another to Greece, 1,200 miles away. The diminutive pipistrelles of Europe also have ranges comparable to the noctule and migrate extraordinary distances.[21]

Similarities between the migratory habits of bats and those of birds are striking. Large groups of bats travel along the same routes as, and often in company with, night-flying birds. Merlin Tuttle, a leading bat conservationist, has sighted migrating groups of Red bats numbering over 100,[22] and mixed "flocks" of red, seminole, hoary, and northern yellow bats have been seen flying with birds.[23] Tuttle has also reported that in the Lake Michigan area during September and October, waves of migrating bats often use passing low-pressure systems to their advantage, possibly attaining speeds of 80 mph with strong following winds (see p. 57).

The navigational abilities of bats, within their more limited migratory ranges, are also no less inspiring than those of birds, though overall, successful homing of displaced bats is generally lower. In various studies, little brown bats released 250 miles from their summer roost, well outside their area of familiarity, have returned over a period of 17 to 22 days, and a female pallid bat displaced in different directions and distances up to 70 miles from its roost site made 8 consecutive homing flights, the longest distance completed in 6 days. But juveniles in similar situations seem to lack homing ability and usually settle for acceptable roost sites near their point of release.[24]

Far less is known about the navigational aids used by adult bats than is the case with birds. Bats are highly regarded, of course, for their extraordinary echolocation capabilities, but as a navigational tool, this trait is probably ineffective. The use of echolocation alone to guide the bat in a migratory flight of many miles would require memorization of an "auditory corridor" of considerable length. Even if this were possible, a more limiting problem is that high-frequency signals emitted by bats drop off rapidly in air, rendering them ineffective at distances greater than 200 to 300 ft.[25] Since migratory bats frequently fly at altitudes greatly exceeding this, some other means of navigation is implicated, and the weight of evidence suggests that it is probably visual. Bats do have good eyesight and use it to their advantage. In a study of spearnosed bats in Trinidad, blindfolded animals released with control groups from a point just 6 miles from their roost circled aimlessly for nearly 7 hours, while the bats that could see headed in the direction of their roost within 30 minutes.[26] This begs the question, however, of what they cue on at night in unfamiliar territory, and few answers have been forthcoming. Sunset glow appears to be important in establishing direction for normal foraging flights, and stellar navigation seems a possibility for longer migratory flights, but evidence in support of either is tentative.

Unlike birds that migrate in autumn to new winter foraging territories in southern climes and then return north to breed, many northern bats migrate solely to hibernate. These movements are often, but not always, latitudinal. The better known bats, like the Mexican free-tailed that summers in the United States, and the noctule of northern Europe, show strongly directional southward or southwestward migration in autumn, as do the tree bats (red, hoary, silver-haired, Seminole, Northern yellow) that hibernate in sheltered but relatively open situations in the southern part of their range.[27] Others, however, particularly the many species of myotis that hibernate in caves, may even migrate northward to large hibernacula, where cooler temperatures and higher humidity are more favorable to hibernation than conditions in their maternity caves or summer roosts. Several myotis, for example, migrate from all parts of southern New England (and elsewhere in the northeast) to an overwintering site in Vermont. Florida populations of the gray bat migrate northward in autumn to caves in northern Alabama, Tennessee, and West Virginia. Fidelity to a particular hibernation cave is virtually 100%, regardless of where the bat spends the summer (of 3,110 gray bats banded in one of the 3 known winter hibernacula used by this species, not a single individual was ever found wintering in any of the other caves in 14 years of study!).[28]

Not surprisingly, the prime physiological requisite for successful overwintering in the bat is also fat accumulation. Unable to switch diet to higher carbohydrate foods like many birds, and, in fact, unable to build fat reserves through hyperphagia as birds do, bats must resort to a different strategy. Instead of overeating, bats underutilize the energy they do obtain. Taking advantage of their ability to go into torpor readily, bats begin to reduce their energy consumption in the fall, well in advance of hibernation, by entering torpor more frequently during the hours in which they are not foraging. In so doing, some small excess of food energy is converted into fat—less than a hundredth of an ounce per day in the fringed myotis, according to one study. Little by little, a quarter-ounce bat can build up a reserve of perhaps 25 to 30% of its fat-free weight—not much compared to the weight gains of migrating birds or other hibernating animals, and not much room for error. Whereas the small pocket mouse, with its cached food stores, may arouse every 3 days or so, each episode lasting several hours, a similar sized bat may sustain no more than a dozen arousals all winter, each lasting only an hour or two (this is why disturbance to bats in caves during hibernation, causing unnecessary arousal, is so threatening). Calculations based on minimum metabolic rates and the caloric value of fat suggest that the big brown bat may carry enough reserve for 190 days, while the smaller myotis species have a maximum hibernation potential of about 165 days. At the lowest end of the scale, the male Western pipistrelle might manage 90 days at best. With such a narrow margin of safety, underprepared juveniles often suffer heavy mortality.[29]

Fall mating is also the rule with hibernating bats, dictated in this case not by a long gestation period, but by the need of females to move to maternity colonies and give birth as early in the spring as possible, usually before males have terminated their hibernation. Furthermore, male bats, unlike larger mammals, do not carry enough reserve energy at the end of hibernation to develop reproductively. So rather than jeopardize juvenile survival, mating occurs prior to or during migration (recalling the mating habitats of caribou), or just before entering the hibernaculum, with copulation most often taking place in flight. Merlin Tuttle recounts observing red bats falling from the sky in pairs as they attempted to mate, and another instance watching two flying together, one repeatedly trying to land on the other and finally uniting while they continued in flight with all four wings flapping! Courtship is unknown, and males will mate with as many females as they can. Sperm is stored in the female reproductive tract over winter, and fertilization takes place as soon as the females emerge from hibernation in the spring.[30] It is just one more unique twist for a mammal that has more than its share of unusual qualities.

III. A TIME TO SOW, A TIME TO REAP

Finding New Ground

&. Autumn is a time for dispersal, a time to stake new territory, a time to sow the seeds of perpetuity. After a summer of parental care, the young of another generation are given their hardest lesson, abandoned into a world of hostilities to make it on their own. For some the separation may follow a period of cohabitation and learning. Yearling bears benefit from a full annual cycle with their mothers before being chased off in their second autumn to find their own place, their own mate, their own winter den; young bull elk are tolerated in the herd for a year before they are evicted by the adult cows to compete for the rights of gender; beaver kits, growing larger and maturing sexually, are ousted from the lodge by their elders at the end of their second summer to seek new waterways and test their own skills. For others, however, the separation comes much earlier in life, with little experience and without any knowledge of the winter that lies ahead. In the abridged world of small mammals, offspring quickly become competition for food and mates—and there is little room for altruism where life expectancy may be measured in months instead of years. Family ties break easily under the pressures of survival. And if the juvenile, pushed by its parent to the periphery of its natal ground, finds only marginal habitat in which to fend for itself, it will have to move on, often through hostile territory where only the fittest—or luckiest—will succeed. But this is the way natural selection works. The need to roam, the need to find new ground, is as fundamental to both plants and animals as is their individual growth and reproduction. It is the final, crucial step in the effort to perpetuate their kind. It is the means by which the whole of the species tracks shifting resource patterns in an ever-changing environment. It is the hope of the future.

While it seems that nothing so important should be left entirely to chance, dispersal is, in fact, something of a lottery. Finding a safe site—a site that satisfies the species requirements for establishment and growth, a site that is adequate of resources and slight of predation, yet not isolated from pollinators or potential mates—is, but for the most specialized of dispersers, largely a blind venture, a matter of chance events on a playing field of ever-changing boundaries and conditions. In terms of gamesmanship, however, it is not the perfect lottery. The moves of individual players are not necessarily random, nor are the odds of winning distributed evenly among all participants. Dispersal abilities differ widely among animals, depending on size, mobility, degree of habitat or food specialization, and response to edge or boundary conditions (willingness, for example, to cross a road, open field, or other alien environment). Similarly, plants differ widely in seed mobility, depending largely on what agents they utilize for transport. The objective is, nonetheless, the same for all players: to get away from your parent, and then get farther if you can.

It is in the fall of the year, particularly, that plants display their ultimate strategies for survival of the species. The geometric beauty of dried seed heads in autumn fields is as purposeful as the flamboyant color and symmetry of spring flowers. Just as flower design tends to optimize pollination success by appealing to a particular agent, so, too, does seed architecture influence dispersal efficiency by specific carriers. While there is no best answer to the challenges and limitations of dispersal, each species has evolved a strategy that suits its particular needs—and the variety of tactics that have turned up in nature would impress the cleverest of inventors. Some plants spread their reproductive investment over a great many propagules of small size, saturating an area with seed in order to increase the likelihood that one or a few will survive. Others gamble more resources on fewer but considerably larger fruits, each with a much higher probability of survival by virtue of the energy stores they contain or their attractiveness to animal dispersers that carry them off to more favorable places. Some plants float their seed on the wind to sail wherever fate will take them; others shatter their pods explosively to throw seed just beyond the shadow of the parent. Some propagules are designed with hooked appendages or sticky secretions to hitchhike on unsuspecting passersby; others have coevolved a nearly complete dependence on just one or two animal species for their dissemination.

Among the most conspicuous and best known of plant dispersal mechanisms are those involving special modifications of seed for wind transport. Is there any child who has never waved a dandelion seedhead or blown at it to watch its tiny parachutes drift away, or plucked at the silk of milkweed pods in the fall to launch those seeds into flight? In these and many similar wind-dispersed species, the structure that forms the plume itself is actually a familiar, though often unrecognized, part of the flower—a modified calyx (the whorl of sepals or lowest tier of flower parts) that remains attached to the seed capsule at maturity. The downy nature of this structure, coupled with a lightweight seed, imparts considerable buoyancy to the propagule, greatly slowing its fall to the ground. Even slight convection currents may then intercept its descent and sweep it upward, now and again carrying it to high altitudes and faraway places. It is a rare seed out of many thousands, however, that travels great distance and lands by chance on a patch of suitable ground. When it does, the successful establishment of a new population is a momentous occasion for the species—but inasmuch as many wind-dispersed plants are low growing, the fate of much seed is to ground out close to the parent. Such is the sweepstakes nature of long-distance dispersal.

Other modifications for wind transport include the development of wing-like appendages on seed capsules. These commonly take the form of a samara, as in the familiar maple and ash seeds, or a winged nutlet, as in many of the conifers, birches, and alders. The samara functions by autogyration, its seed weighting one end of the wing and causing it to spin like a helicopter blade as it falls. This slows its rate of descent, thereby increasing the amount of drift that might occur away from the parent plant. With a good breeze and an unobstructed fall, a samara released from high in the tree might stray horizontally 150 ft or more,[1] well out of the shadow of its parent. The plane-wing form of pine, birch, and alder seeds imparts a much more variable flight pattern, usually resulting in a fluttering, spiraling, or rolling descent. Unlike the more buoyant plumed seeds, samaras and other winged forms typical of many north temperate and boreal tree species generally travel only tens of feet from the parent, particularly when released within a forest stand.[2] Late-season fallout onto snow may result in added movement by wind, but longer transport is usually accomplished only with the help of animal removal and subsequent caching or dropping of these seeds some distance away (see next section).

The simplest of all modifications for wind dispersal involves the production of minute propagules, sometimes termed "dust seeds," that float readily in the slightest of air currents without any other morphological help. These may weigh no more than 0.01 mg, and in some plants may actually weigh less than 0.004 mg (approaching one millionth of an ounce!).[3] Species producing dust seeds are found in many genera, including several rushes (*Juncus*), stonecrops (*Sedum*), bellflowers (*Campanula*), wintergreens *(Pyrola)*, saxifrages (*Saxifraga*), and, perhaps most notably, the orchids. In some cases the minute size of the seed is augmented by a

balloon-like veil consisting of large empty cells with air-filled intercellular spaces, a papery remnant of the outer integument that surrounded
the embryo and that now gives the propagule extra buoyancy.[4] Some of
the orchids, wintergreens, and saxifrages possess this feature, as do the
seeds of pitcher plants and sundews found growing in acid bogs.

Included among these floating "dust seeds" must be counted the innumerable minute spores produced by nonflowering plants—the many
ferns, horsetails, lycopods (club mosses, ground pines, or running
pines), and the true mosses as well as the fungi. Spores are reproductive
propagules, normally unicellular, capable of growing into mature plants
without fusion with another cell, and are produced within specialized
structures on nonflowering plants known as sporangia.[5] The sporangia
are visible as small dots or clusters on the undersides of fern leaves (or in
some ferns as separate stalks or fertile fronds), or as tiny yellow seed-like
structures in the leaf axils or fruiting stalks of lycopods. In mosses they
are contained within capsules on a slender stem that rises above the
feathery ground cover. When mature and ready to dehist, the sporangia
break open to release small clouds of spores—countless millions of
them—that rise like wisps of smoke on imperceptible air currents.
Walking through carpets of lycopods and mosses in autumn woods
often leaves boots yellow with spore dust.

One of the more creative forms of wind dispersal, noteworthy here if
only for its uniqueness, is the tumbling habit of certain plants of open
country. Tumbling is a mechanism developed primarily among a few
annuals of steppes, deserts, and prairies, mostly of Eurasia, whose seeds
are dispersed only upon rolling. The "tumbleweeds," exemplified by
Russian thistle (which has become naturalized throughout the western
United States), usually grow in the form of densely branched, compact
plants, as wide as they are tall and becoming more or less spherical in
shape as they mature and dry. Rather than releasing seed in place at the
end of the growing season, however, the whole plant detaches from its
base via an abscission layer where it is weakly held. Driven by the wind,
the tumbleweed lurches and bounds across the open landscape, spilling
seed as it goes.[6]

While a number of weedy plants like dandelion and Russian thistle
have managed to spread throughout the northern hemisphere, the overall effectiveness of wind dispersal is far less than would be suggested by
the success of these species alone (much of their invasiveness can be
attributed to assistance from humans, coupled with opportunistic seedling establishment traits and a broad tolerance to stressful conditions,

particularly those associated with disturbed sites). In reality, wind dispersal without subsequent redistribution of seed appears the least effective mechanism in reaching appropriate sites, with a high percentage of seed missing its target, lodging in vegetation instead of soil, falling out over water or other inappropriate places, and eventually perishing. While a rare outcome can have great significance in the distribution of a plant species, wind dispersal is the gambler's approach, throwing seed out with hope and a promise, saturating an area for a chance success.

ONE MILKWEED WITH FAITH

[T]he Asclepias [milkweed pods] now point upward . . . already bursting. I release some seeds with the long, fine silk attached. The fine threads fly apart at once, open with a spring, and then ray themselves out into a hemispherical form, each thread freeing itself from it neighbor and all reflecting prismatic or rainbow tints. The seeds, besides, are furnished with wings, which plainly keep them steady and prevent their whirling round, I let one go, and it rises slowly and uncertainly at first, now driven this way, then that, by currents which I cannot perceive, and I fear it will make shipwreck against the neighboring wood; but no, as it approaches it, it surely rises above it, and then, feeling the strong north wind, it is borne off rapidly in the opposite direction, ever rising higher and higher and tossing and heaved about with every fluctuation of the air, till, at a hundred feet above the earth and fifty rods off, steering south, I lose sight of it. How many myriads go sailing away at this season, high over hill and meadow and river, on various tacks until the wind lulls, to plant their race in new localities, who can tell how many miles distant! And for this end these silken streamers have been perfecting all summer, snugly packed in this light chest,—a perfect adaptation to this end, a prophecy not only of the fall but of future springs. Who could believe in prophecies of Daniel or of Miller that the world would end this summer, while one milkweed with faith matured its seeds?

—HENRY DAVID THOREAU, 1851[7]

One alternative to random wind dispersal is the packaging of seed within an attractive fruit which becomes, in essence, a reward for dispersal services rendered to the plant by frugivorous (fruit-eating) animals.

The animal is offered the nutritious pulp in exchange for scattering un-digested seed as it roams. While still a chance proposition, animal dis-persal offers a somewhat more directed distribution of seed, particularly where specific carriers are involved, because animals tend to move between like habitats and are, therefore, more apt to transport seed from one suitable site (the parent site) to a similar one elsewhere. Both the pattern and distance of dispersal by frugivores are dependent, however, on the ranging behavior of the animal and on its food processing char-acteristics. Two examples will serve: While birds are capable of moving long distances after ingesting fruits, their small size and relatively short digestive tracts require fairly rapid processing of food. Birds often sep-arate pulp from seed in their upper gut and regurgitate the larger seeds within a matter of minutes, compared to an hour or more for defecation of undigested seed. Thus, internal transport by birds is more likely to be of a local nature. In contrast, larger, wide-ranging omnivores like bears and coyotes that ingest considerable amounts of fruits in the fall process the pulp longer and concentrate more seed in small clumps, distributed over large territories.[8]

In some cases seeds, by virtue of their placement rather than their "packaging," are ingested incidentally to the pursuit of an unrelated re-ward such as the leaves of grasses. The result is the same, however: A few seeds survive passage through the animal's digestive system and are voided with the animal's feces, wrapped in their own packet of fertilizer to establish the species anew some distance from the parent plant. Though the distinction between seed disperser and seed predator be-comes rather hazy sometimes, as the seed itself is often the target of consumption (with the objective being to digest it for its energy), even if only a small percentage of ingested seeds—say 5%—survives the chemical and abrasive action of the animal's gut, the number of viable propagules disseminated over the period of seed ripening may be im-pressive. With some sacrifice of seed necessary to insure the dispersal of others, natural selection seems to have found the balance: Variation in the digestibility of seed is accomplished through variation in hardness of seed coat within plant populations.[9] Natural selection seems also to have favored bright packaging to attract attention to the fruit, though the argument here is equivocal.

> The new Botany teaches that the flowers have color and perfume to attract the insects to aid in their fertilization ... Is it not equally true that the high color of most fruits is to attract some hungry creature to come and eat them and thus

scatter the seeds? From the dwarf cornel, or bunch-berry, in the woods, to the red thorn in the fields, every fruit-bearing plant and shrub and tree seems to advertise itself to the passer-by in its bright hues. Apparently there is no other use to the plant of the fleshy pericarp than to serve as a bait or wage for some animal to come and sow its seed. Why, then, should it not take on these alluring colors to help along this end? And yet there comes the thought, may not this scarlet and gold of the berries and tree fruits be the inevitable result of the chemistry of ripening, as it is with the autumn foliage? What benefit to the tree, directly or indirectly, is all this wealth of color in the autumn?...The cherry-birds find the pale ox-hearts as readily as they do the brilliant Murillo, and the dull blue cedar berries and the duller drupes of the lotus are not concealed from them nor from the robins.

JOHN BURROUGHS, 1905[10]

In yet another evolutionary twist, some plants have developed propagules that disperse by adhesion or "hitchhiking" on unsuspecting animals without returning any favors. In this case the plant neither advertises its fruit to attract the animal, nor offers it any reward for carrying the seed, but rather depends on chance encounter and then clings to the animal by some elaborate modification of the propagule. The animal is usually unaware that it is carrying the seed until some time after it has picked it up. Clinging appendages normally develop from the outer fruit wall (pericarp) rather than from the seed itself, and in some plant families bear close relationship to wind dispersal features, suggesting that the latter was ancestral to adhesive dispersal.[11]

The majority of adaptations for this mode of transport consist of barbs or hooks, such as are found on the familiar beggar's-ticks and burdock. Some are so refined as to be aimed at specific animals. The sweeping hooks of the devil's claw that wrap around the hooves of grazing ungulates in the American southwest may be one of the more outstanding examples of host-specific adaptation. But even seemingly ordinary burrs are often more effective in clinging to the coats of some animals than to others, depending on the finer details of their structure. A smaller number of adhesive propagules cling by sticky or viscid outgrowths of the seed, some of which may be contained within a fleshy fruit. The viscid seed of the mistletoes, for example, might thus be carried considerably farther by adhesion to a bird's feathers than would occur solely by defecation of the undigested seed.[12]

Though the rate at which adhesive seeds are picked up by animals tends to be quite low, once attached, the potential for long-distance transport is higher than for any other mode of dispersal. For most seed, the ride ends only when groomed from the animal's pelage or plumage. Large burrs in conspicuous places may be discovered and removed relatively quickly, but small propagules in hard-to-reach locations like an animal's back may remain attached for long periods of time and, with some animals, travel great distances. This is perhaps best illustrated by studying the plant geography of remote islands in relation to dispersal mechanisms of the established vegetation. On the Cocos Islands in the southern Indian Ocean, for example, situated 950 miles from Sumatra, the nearest land mass, only those plants that disperse by ocean drift and adhesive fruits are found.[13] And on Macquarie Island, 600 miles out to sea between New Zealand and Antarctica, all 35 plant species found are animal dispersed, most by adhesive propagules.[14] Analysis only of existing vegetation does not reveal the nature of seed that might have arrived by other means but failed to establish, but it does indicate the effectiveness of adhesive fruits in long-distance dispersal.

While it is the architecturally elaborate that tends to attract most of our attention, the majority of hitchhiking seeds have no special adaptations at all, but simply cling with mud to the feet of animals. Charles Darwin made several observations to this effect, noting once that a "Mr. Swaysland, of Brighton, who during the last forty years has paid close attention to our migratory birds, informs me that he has often shot wagtails, wheatears, and windchats, on their first arrival on our shores, before they had alighted; and he has several times noticed little cakes of earth attached to their feet." Darwin then related an instance in which he had been sent the leg of a red-legged partridge "with a ball of hard earth adhering to it, and weighing six and a half ounces. The earth had been kept for three years, but when broken, watered and placed under a bell glass, no less than 82 plants sprung from it: these consisted of 12 monocotyledons, including the common oat, and at least one kind of grass, and of 70 dicotyledons, which consisted, judging from the young leaves, of at least three distinct species." Darwin went on to speculate, "With such fact before us, can we doubt that the many birds which are annually blown by gales across great spaces of ocean, and which annually migrate . . . must occasionally transport a few seeds embedded in dirt adhering to their feet or beaks?"[15] It is entirely possible that for plants lacking any obvious dispersal mechanisms, this may be the most common mode of seed dissemination.

Harvesting the Future

The general belief among country-people that the jay hoards up nuts for winter use has probably some foundation in fact, though one is at a loss to know where he could place his stores so that they would not be pilfered by the mice and the squirrels. An old hunter told me he had seen jays secreting beechnuts in a knothole in a tree. Probably a red squirrel saw them too, and laughed behind his tail.

JOHN BURROUGHS, 1891[16]

It is an incontrovertible fact of plant life that much of its seed production, whether by design or not, will go directly into the mouths of hungry animals. To the plant this is part of the cost of doing business. If it is to reproduce successfully it must provide sufficient seed to satisfy the demands of consumers and still have enough left over to assure some chance of seedling establishment. But plants long ago settled their differences with the herbivores, having coevolved with them since before the beginning of the Cenozoic, some 70 million years ago. In fact, the seed harvesters themselves have played a crucial role in shaping plant–herbivore relationships, culling out of the plant population by periodic overharvesting those genetic traits that result in low seed production. And sometimes, as we have already seen, the harvester becomes a partner, witting or otherwise, in the dissemination of seed, paying back some of the added cost of seed production through enhanced germination success. In some cases even, there has evolved between plant and harvester a relationship of mutual dependency that seems to assure the survival of both.

One of the most remarkable stories of codependency between plants and animals in the temperate latitudes is played out during the fall of the year in the soft-pine forests of western North America—most notably within the sprawling piñon-juniper woodlands of mid elevations and the whitebark and limber pine stands at timberline—and across 4,000 miles of almost uninterrupted stone pine forests in Eurasia. It is a story of autumn seed harvest, of winter food caches, and spring birth—a story of seed production tailored to the needs of a guild of birds, and of birds shaping the evolution of seeds in a group of pines. First, the birds.

Several members of the Corvidae family, which includes crows, ravens, jays, and magpies, are known to cache food. For some, the cache is little more than an insurance policy, a small hedge against life's uncertainties, but in the western United States two of these birds, the piñon jay and the Clark's nutcracker (also a large jay), have become almost entirely dependent on scattered hoards of pine seed for their winter sustenance. Throughout the fall, these birds roam far in search of bountiful cone crops (conifers are noted for their sporadic cone production) and, once found, harvest seed by the tens of thousands.

Clark's nutcrackers (named after Captain William Clark for his description of the bird in the Bitterroot Mountains in 1805[2]) are birds of the high pine forests, remaining near timberline throughout frozen, windswept winters, where any semblance of food would appear to have gone the way of summer. They share this habitat most frequently with whitebark and limber pines, species that would seem equally luckless in finding the necessary resources to hang on, let alone to reproduce successfully in such a difficult environment. But in the large, awkward seeds of these pines the nutcracker has found a fortune—and in the bird, the trees have found a solution to at least one of their problems.

The Clark's nutcracker is particularly well adapted to relieving these trees of their seeds. With a bill like a pry bar, it can open the scales of any cone, no matter how tightly closed, and does so with great energy and skill. It is equally well equipped to expropriate seed once it gets into the cones. Having a sublingual pouch—a distensible sac positioned beneath its tongue—that is unique in the bird world, the nutcracker can transport 90 or more large nut-like seeds at once, or twice that many smaller ones, for a payload of about one fifth of its own weight.[3] One nutcracker laying in winter stores may transport seed up to 10 miles, and may stash as many as 100,000 seeds in a season.[4] And most of these it secrets off to windswept ridges or sunny south-facing slopes where thin snow cover or early melt permits their recovery throughout the winter and spring. There the birds "plant" seed in shallow soil, sometimes singly, but usually in groups of two to a dozen or more, by first probing the soil with the bill and then loading a seed into the bill (from the sublingual pouch) and jamming it into the loosened soil one at a time. The cluster of seeds is then covered loosely, occasionally with a small twig, pebble, or empty cone placed over it. Several caches may be made in a suitable locality by nutcrackers working singly or in small groups, but while the caching ground may be communal property, individual hoards apparently are not shared.[5]

The piñon jay, though a smaller bird than the nutcracker, is also able to transport large numbers of seed in its expandable esophagus (up to 15% of its own weight—still far more than other jays[6]) and is equally committed to the task. What this jay lacks in size it makes up in numbers, frequently traveling in large cohesive flocks, sometimes exceeding 200 birds, that harvest and move seed by the hundreds of thousands. Like the nutcrackers, they begin harvesting the year's new crop when the cones are still green, prying cone scales open with their strong bills and feeding on the still-maturing seed; once the cones dry and open by themselves, caching begins in earnest. In noisy waves the jays sweep through productive stands, each bird selectively gathering from the ground or tree only the fully developed seed, rapidly stashing a dozen or more into its expandable esophagus, and then whisking off to their traditional nesting territory where it caches the seed in inestimable numbers, singly or in small clusters. With each bird in the flock capable of gathering 10,000 to 20,000 piñon nuts in a fall season,[7] the ground in their caching area soon becomes permeated with pine seeds—seeds to see them through the winter, through courtship and breeding, through the rearing of another generation.

While nutcrackers may cache seeds from a number of different pine species, principally those with larger wingless seeds, piñon jays are much more selective, rarely harvesting seed other than from the piñon pine—and for their service, the piñon has returned the favor. The large seed, often referred to as a "nut" for it size, is packed with food value— 14% protein, 18% carbohydrate, and 65% fat. Put in human terms, a pound of these seeds shelled provides more calories than a pound of chocolate and almost as much as a pound of butter, with more utilizable protein than any other commercial nut but cashews. All 20 amino acids are present in the piñon, and 85% of the fats are unsaturated oleate, linoleate, and linolenate.[8] These statistics are not wasted on the bird, either. The piñon jay has found in these seeds an extremely concentrated high-protein, high-fat, energy source in an unarmored cone that opens wide over a several-month period of ripening for effortless harvesting.

What does the piñon gain for being so attractive to these seed-hoarding birds? Except for certain subspecies in the Sierra Madre of northern Mexico that still produce small winged seeds, piñon pines have largely abandoned wind dissemination in favor of more directed dispersal by birds. Without the jays the now large, wingless piñon nut would merely drop to the ground beneath the parent to desiccate on the dry surface litter. Throughout most of the range of piñons, rainfall is scant and the only chance a seed has of germinating and establishing is through serendipitous burial in a moist mineral soil—hence the advantage of partnering with the jays. A large number of high-food-value seeds are exchanged for the placement of few extras in locations suitable for gemination. The jay is rewarded, and the uncertainty of chance dispersal is traded for the greater probability that the bird will not recover all the seeds it buries. And the jays have undoubtedly contributed to the evolutionary success of this arrangement through their careful selection of seed for harvest. Rather than indiscriminately collecting all seed encountered, the piñon jay assesses the value of each nut, "weighing" or clicking it in its bill to be sure it is fully developed, rather than investing energy in a seed that is unfertilized, diseased, or occupied by an insect larva.[9] Not only does this have survival value for the jay, but by burying only the larger seeds the jays are unwittingly selecting for better trees in the future, increasing the likelihood that any germinating progeny will one day itself produce large seed.

. . .

The manner of food caching described here for nutcrackers and jays is termed "scatter hoarding" and is typical of a number of animals that store food in autumn for use when resources later become scarce. Indeed, much of the fall activity of small mammals, even of some carnivores, centers on caching some part of the bounty of the season for future use. And much has been written about the importance of food hoarding animals to the successful regeneration of certain plant species. Gray squirrels, for example, are thought to be instrumental in the dispersal of some of the large nut trees of eastern deciduous forests. But implicit in all behavior that involves scatter hoarding is the assumption that the animal doing the caching has a reasonable chance of recovering the food item later even when snow covers the site. The question as to how this is accomplished, especially for birds that bury several thousand food items and lack a well-developed sense of smell, is in itself a fascinating story.

> I have found untouched hoards of acorns that squirrels had gathered, but left because the shell-barks also stored had proved sufficient. And yet I have seen squirrels bury them with care, as though foreseeing their needs, and planting an oak

for their indefinitely great grandchildren. I do not suppose a squirrel proposes to disinter the nuts it hides singly in the ground and use them as food. A mammal with such an extraordinary memory would soon cease to be a mere mammal by reason of it. It is, perhaps, as hard to believe that it plants the acorn that a tree may grow. Why it does it is a problem yet to be solved.

CHARLES ABBOTT, 1889[10]

While studying the hoarding behavior of captive birds in large enclosures, investigators have repeatedly found that unless a bird cached a food item itself, its chances of finding a hidden stash were rarely better than for random search. Seldom did nutcrackers or jays find food hidden by others, even though they often passed close by another's hoard. Yet the percentage recovery of a bird's own caches, either in the field or under experimental conditions, is often very high—up to 99% in some cases.[11] All observations suggest that birds rely on some visual cue, rather than sense of smell, to recover their caches.

It has already been noted that when nutcrackers and jays bury seeds, they often rake over the soil and then drop an object like a pebble on top of the cache site. Yet when Stephen Vander Wall rearranged all soil features in one half of an experimental enclosure after nutcrackers had buried seed, the birds still recovered their caches efficiently. It was only when he changed the spatial relationships of larger "landmarks" that the birds became confused. After allowing birds to cache food a second time, Vander Wall removed his subjects and extended one end of his oval arena in the direction parallel to its long axis by just 8 inches, and then displaced four large rocks previously situated at that end of the enclosure by exactly the same distance and direction. When the nutcrackers reentered the arena and sought out their caches, recovery was nearly complete at the unaltered end of the enclosure. Where the perimeter had been extended and the rocks moved, however, the birds missed their caches by a mean distance of 8.1 inches! They probed for their caches in almost exactly the same position relative to the fixed objects and perimeter of the enclosure at the time they buried the seed, indicating clear remembrance of where they had cached food, based on spacial reference points rather than marking the site with visual cues—and experiments like this have found memory dependable to at least 31 days.[12]

If the challenge of recalling information on the locations of thousands of caches placed over a period of several months were not difficult enough, nutcrackers and jays have been reported to recover food beneath snow sufficiently deep to obliterate all ground surface features, and with an 80% success rate.[13] Even in deep snow these birds are able to pinpoint a cache location based on only a few elements of a remembered field of view and dig to that point, judging angle and distance correctly when they are unable to see the visual cues above the snow. Eurasian nutcrackers in Sweden and the Soviet Union were observed to recover food under 18 to 34 inches of snow![14]

Surprisingly, birds are not the only food hoarders to rely on spatial memory for recovery of caches. The same may be true of mammals as well, in spite of their superior olfactory senses. While there is no question that mammals are better able than birds to search out food by smell, numerous studies suggest that the owner of a cache is far more likely to find it than is an uninformed passerby. When a semitame red fox, for example, was allowed to cache mice along the periphery of a field, it was able a day later to walk directly to 48 of 50 caches without any exploratory behavior, suggesting it knew exactly where each was buried. However, in a subsequent repeat of the experiment, the fox was again allowed to cache mice, but after its removal from the area an investigator buried additional mice within 10 ft of each of the fox's caches. When allowed to return, the food-deprived fox again went straight to its own hoards, but accidentally found only 2 of 20 added mice, which it discovered by smell when it nearly stepped on them. And when later the fox's own caches were moved by as little as 3 ft from their original location, the recovery rate dropped to 25%. In this case, the fox dug extensively around the empty sites before finally sniffing around to discover the recached mice. Captive foxes were very poor at discovering each other's caches, too. After one fox was permitted to cache food along a trail, two other foxes were allowed to walk the same trail, neither discovering any of the caches. Later, the cache owner recovered each of them.[15]

Similar studies with other mammals have corroborated these results. Gray squirrels, for example, are consistently better at finding their own caches than supplemental ones placed in the same area by investigators. And when red squirrels hide fungi high in the tops of conifers they unerringly return to the right tree to recover them, though the fungi are out of sight and almost certainly beyond olfactory range.[16] Observations like these suggest again that caches go mostly to the cacher. While there is no discounting the advantage of good olfactory senses, when it comes to food recovery it appears that the nose works better in conjunction with memory.

The big question, then, as all this relates to autumn seed dispersal and the reproductive strategy of trees, is, can jays and nutcrackers remember the location of every cache they make from September through December (and remember 3 months into winter which of the caches they have already recovered so that they don't waste time revisiting them, only to find them empty)? Not likely. If a nutcracker hoards 100,000 seeds in a season, burying an average of 10 seeds per cache (to keep the numbers round), that's 10,000 individual caches. But a nutcracker doesn't have to remember all locations, because in a good year it will hoard the equivalent of 2 to 3 times its likely energetic needs.[17] Thus a 50% recovery rate, the lowest estimate in the literature, will be adequate for most years— which leaves a lot of seed left over for another generation of conifers.

The nature of scatter hoarding is such that, from the tree's perspective, investment in large attractive seeds has a reasonable chance of paying off, provided that the seeds are not continually overharvested (all seeds recovered) or that postdispersal mortality of seeds is not inordinately great. But nature seems to have taken care of this matter, too, through the sporadic production of cone or mast crops. With wide temporal and spatial variation in seed production—that is, with the production of good seed crops only once every 2 to 5 years and with little in between—it is difficult for potential harvesters to track the seed resource and, hence, overconsume it. At the same time, the high degree of regional synchrony in cone or mast crops assures that periodically a large excess of seed will be cached, leaving many to germinate. And for long-lived trees the successful establishment of only a few seedlings per decade is sufficient to keep the species going.

. . .

It is with larder hoarders that the relationship between plant and harvester breaks down. Larder hoarding involves the concentration of food stores in one or a few centralized locations, rather than in many small caches scattered throughout a territory, and is typically practiced by animals that are confined to a den or whose foraging mobility is otherwise restricted during winter. Some larder hoarders have relatively little effect on seed dispersal, as they mostly accumulate vegetative material for later use when its availability becomes limited. Beavers and pikas are good examples, each stockpiling woody shoots and leafy stems, respectively, for consumption under ice and snow where the animals remain active during winter (discussed later). But others, most notably the winter-dormant ground squirrels and chipmunks, cache large quantities of seed in their dens to supplement fat reserves utilized during periodic arousals or early emergence from hibernation (p. 86).

While these mammals also store dried vegetation, fungi, and fruits on occasion, it is the seed they depend on most—and the granaries of larder hoarders sometimes reach impressive proportions. The burrow of an eastern chipmunk was once described by John James Audubon and a collaborator as containing within several chambers one gill (approximately ½ cup) of mixed wheat and buckwheat seeds, a quart of hazelnuts, 2 quarts of unmixed buckwheat seed, and a peck (8 quarts) of acorns.[18] Many similar accounts have since been published.

Larder hoarding is also an efficient strategy for winter-active animals that are highly territorial and able to defend centralized food stores. One of the better known of these is the ubiquitous red squirrel, an energetic, seed-eating specialist of temperate and northern coniferous forests, whose storage middens are often conspicuous features of the forest floor. Dependent in large measure on cached food for its high caloric demands in winter, the red squirrel begins harvesting in earnest as soon as the cones of spruce, fir, and pine mature in late summer or early fall. With bird-like agility ("Give him wings and he would outfly any bird in the woods," wrote John Muir), it scrambles through tree tops felling cones by nipping at their bases, sometimes clipping the whole branch tip on which they hang and dropping several at a time.

> He threads the tasseled branches of the Pines, stirring their needles like a rustling breeze; now shooting across openings in arrowy lines; now launching in curves, glinting deftly from side to side in sudden zigzags, and swirling in giddy loops and spirals around the knotty trunks; getting into what seem to be the most impossible situations without sense of danger; now on his haunches, now on his head; yet ever graceful, and punctuating his most irrepressible outbursts of energy with little dots and dashes of perfect repose. He is, without exception, the wildest animal I ever saw—a fiery, sputtering little bolt of life, luxuriating in quick oxygen and the woods' best juices.   JOHN MUIR, 1911[19]

Tree by tree the squirrel gathers the year's cone crop into its larder. If the cones are to provide useable energy over a several-month period, however, it is essential that they be stored in such a way as to prevent the scales from opening and exposing their seed. A young squirrel in new territory will bury cones in clusters, sometimes as many as 20 or 30 at a time, in holes dug 6 inches or more into the damp forest floor beneath a dense stand of conifers. In the cool shade the cones will remain closed, though eventually some quantity of seed will be lost to fungi and other soil decomposers. But if the squirrel is in prime territory occupied by generations before it, the cones are cached in the already established middens. A huge scrap pile of debris accumulated from years of harvesting, burying, digging up, and shucking cone scales in the same place, the midden provides an ideal organic mulch for keeping the larder cool and moist, with minimal losses. Buried in the midden, the cones of spruce and fir may keep without opening for 2 years or more.

Other cones may require different handling, however. The serotinous cones of fire-adapted trees are designed to remain closed tightly until the heat of a forest fire releases their seed. These normally hang on the trees for years, where exposure to drying sun and wind helps preserve them. When harvested by squirrels, then, the serotinous cones are heaped in neat piles of 30 or more on the ground surface, usually at the base of a tree, under a deadfall, or in a hollowed out stump. Burial in moist soil in this case causes the cones to open, soon spoiling the seed. Remarkably, a squirrel occupying a territory where both types of cones are found seems to know exactly how to handle each.

It is little wonder that red squirrels harvest and store cones in the fall as if their lives depended on it. The energy needs of a single squirrel during winter in a temperate region have been estimated at approximately 15,000 kilocalories,[20] and this is supplied almost entirely from conifer seeds. While tree species may vary widely in the caloric value of their seeds, as well as in number of seeds produced per cone, the total required by the squirrel on a daily or seasonal basis is invariably great. In the Rocky Mountains, for example, Engelmann spruce yields about 15 calories per seed and produces on average 180 seeds per cone. Douglas fir, on the other hand, packs 57 calories of energy into each seed but only produces about 45 seeds per cone.[21] In these two examples the energy yield is about the same—approximately 2,500 calories per cone—and in both cases a squirrel will need about 5,000 cones to get through the winter. Lodgepole pine in the same region provides only about a third as much energy per cone as the spruce or Douglas fir, but storage of an entire winter's supply is unnecessary because the cones here are serotinous, retaining their seeds and remaining available on the tree. As winters get longer and colder, needs, of course, increase. Red squirrels in the white spruce forests of interior Alaska may cache as many as 12,000 to 16,000 cones during a good year, accumulating enough unused cones in the midden over time to get through a year of crop failure.[22] While red squirrels may also feed on lichens, cached fungi or other dried fruits, the inner bark of twigs, or occasional carrion, it is doubtful that they would survive winter throughout much of their range without such a larder of cones.[23]

From the plant's perspective there is a significant difference between larder and scatter hoarding: With the latter, seed is widely dispersed with reasonable likelihood that a few will end up at a favorable site and escape later notice, whereas larder hoarding tends to move seed to locations unsuitable for germination or seedling establishment, either because the cache is located in an inappropriate place (usually too far below ground) or seedling competition is excessive from the higher concentration of seeds. Furthermore, the larder hoarder is much more likely than the scatter hoarder to consume all seeds in its cache. Thus, in the scheme to disseminate seed by recruiting the services of animals, larder hoarders appear to be the monkeywrench. But evolution has its ways, and wherever interactions involving two species are consistently strong, natural selection, given enough time, tends to favor traits that permit coexistence.

In the face of extreme seed predation by squirrels, evolution might have gone in one of two different directions, favoring either an increase in seed production to compensate for losses, or greater protection of cones against seed robbery. In the lodgepole pine forests of the Rocky Mountains, evolution seems to be leaning toward the latter option. Here the red squirrel is acting as a major selection agent (a dominant force shaping the reproductive success of the tree species) by preferentially harvesting cones of trees that are the easiest to open and yield the most seed. This leaves trees with tough cones—cones with sharply prickled scales, with a thick resinous coating that keeps the scales tightly closed, and with fewer seeds for the trouble of getting into them—to spread their genetic traits. The result is a seed dispersal strategy favoring serotinous cones that remain on the tree for years, holding their seed tightly against would-be robbers. Cone scales open only when the heat of a ground fire softens their binding resin and releases them into the wind to fall on mineral seed beds that have been cleared of their organic mulch by the burn.

As one measure of the "toughness" of these lodgepole pine cones, only 3% of the total energy invested in the reproductive structure is actually seed, compared to 17 to 46% for the bird dispersed whitebark pine.[24] For its part in shaping the evolution of a tree, though, the squirrel is also changing in anatomy, slowly evolving a stronger jaw musculature to compensate for the tougher cones. In the Sierra Nevada where the relationship between the Douglas squirrel (a subspecies of red squirrel) and lodgepole pine is not as strong, the trees are not serotinous and the squirrel has not developed the jaw musculature of its Rocky Mountain cousin.[25]

. . .

Food hoarding is fundamental to the survival of a great many animals of strongly seasonal environments. Stephen Vander Wall in his exhaustive treatise on the subject[26] discusses no fewer than 284 species of bird, mammal, and insect that stockpile food to one degree or another against the possibility of future shortage. In temperate and northern latitudes the majority of them, not surprisingly, do it solely, or intensify their effort at it, in the fall of the year. Even animals like deer that we do not think of as food hoarders subsidize much of their winter energy demands with fat reserves accumulated in autumn—food storage of another, but no less critical, kind.

For many animals the constraints of winter foraging are clear, the necessity of autumn food caching obvious. Consider the beaver as just one, albeit perhaps extreme, example. Adapted solely to aquatic habitats and an obligate feeder on the inner bark of woody plant species (only occasionally supplementing its diet with tuberous roots of aquatic plants), this animal faces a particularly difficult dilemma during winter. Except under the rarest of circumstances, it spends winter locked under ice, completely isolated from its usual food source. Yet it does not hibernate. So the beaver stockpiles food, clipping up to several hundred saplings or smaller branches from felled trees and either anchoring them in the bottom mud (where an underwater heap nearly as large as the lodge itself may eventually be accumulated), or building a raft of them (where it has been suggested that favored plant species are placed at the bottom of the raft so that they remain accessible when the top is frozen into the ice[27]).

Curiously, however, beavers almost never store enough to satisfy all their winter needs, even though woody plants may be readily available and storage space unrestricted. One exceptionally detailed study of food caches for several colonies in Canada revealed that two out of five were barely sufficient for the number of beavers in the colony (recent energy budget analyses suggest they might have been more inadequate than first thought), and the other three fell far short of meeting the needs of the colony. Yet all survived the winter and population numbers were stable over the long term, suggesting resources were adequate year round. Further studies eventually reconciled the discrepancy, revealing that, for the most part, only the kits used the food cache and that the adults wintered "in relative poverty," subsidizing their energy needs from fat reserves stored in the tail.[28]

Studies like this reveal an important aspect of food hoarding that is easily lost in the details of caching: Seldom is food hoarding by itself, as a preparation against winter energy deficits, sufficient to secure the future of the animal. Adult beaver do not simply subsidize their energy needs from fat stores while leaving the food cache to the juveniles. They employ a suite of mechanisms for reducing those needs, including increasing insulation, lowering body temperature, and reducing peripheral blood circulation.[29] There are two sides to every energy budget, and almost invariably food hoarding behavior is tightly coupled with critical energy-conserving adaptations.

How handsome every one of these leaves . . . soon to turn to mould! Not merely a matted mass of fibers like a sheet of paper, but a perfect organism and system in itself, so that no mortal has ever yet discerned or explored its beauty.

—Henry David Thoreau, 1856

The landscape looked singularly clean and pure and dry, the air like a pure glass being laid over the picture, the trees so tidy and stripped of their leaves; the meadow and pastures clothed with clean, dry grass, looked as if they had been swept; ice on the water and winter in the air but yet not a particle of snow on the ground. The woods, divested in great part of their leaves, are being ventilated. It is the season of perfect works, of hard, tough, ripe twigs, not of tender buds and leaves. The leaves have made their wood, and a myriad new withes stand up all around, pointing to the sky . . .

I experience such an interior comfort, far removed from the sense of cold as if the thin atmosphere were rarefied by heat, were the medium of invisible flames, as if the whole landscape were one great hearthside, that where the shrub-oak leaves rustle on the hillside, I seem to hear a crackling fire and see the pure flames, and I wonder that the dry leaves do not blaze into yellow flames.

—Henry David Thoreau, *1850*

Leaves in varying degrees of decomposition fall to the ground with fungi already colonizing their surfaces (p. 128).

Fruit of the autumn leaf harvest, nutrient-cycling fungi are often stimulated by fall rains to produce mushrooms (p. 134).

Deciduous and evergreen—two answers to the same problem. The aspens, though, hedge their losses with hidden chlorophyll (p. 22).

The term "Indian summer" originated in colonial New England, but the conditions it described may no longer exist (p. 122).

Waiting for snow, Coal Creek, Colorado.

Winter food cache ready, the fresh branches piled underwater (left, beside lodge) are mostly for the beaver kits. Adults will fast (p. 113).

Within limits set by shortening days and decreasing temperatures, genetically distinct aspen clones each turn on their own schedule (p. 19).

In an endless annual cycle of senescence and renewal, buds of leafless aspen await a new year and another season of change.

Season of the Vole

❧ While blazing fall colors herald sweeping change in the aspen and maple wood, no greater transformation could be imagined than that which is occurring with scarcely any notice in the diminutive world of voles. Secretive and often overlooked, these widespread creatures of grassy fields and forest litter will go through more ecological adjustments in the space of a few short autumn weeks than at any other time since birth, for their whole universe is about to change dramatically.

Voles are found throughout temperate and northern latitudes, exploiting a range of environments from prairie to tundra, but almost all with a common denominator—they inhabit highly seasonal environments that are normally cold in winter. Small and short-furred, voles would seem, from a purely physical point of view, to have little in their favor when it comes to conserving heat. Yet they remain active year-round, even in regions of deep snow where they live—and breed—entirely within the confines of the subnivean environment beneath the snow pack. They are, in fact, the very model of adaptive finesse, combining virtually all the energy-balancing tricks in the mammalian world, short of deep hibernation or southward migration, to beat the cold. And fall is their telling season, for the preparatory changes, behavioral as well as physiological, that they undergo at this time are indeed profound.

We have already glimpsed a few of the adaptive strategies by which some mammals ready themselves for winter—the color changes; the added insulation; the accumulation of fat reserves—but in our discussion of hibernators (p. 83), we saw that one of the most effective means of conserving energy in the cold is to reduce the temperature difference between the body and its surroundings. There are several ways, besides hibernating, of approaching this problem, however, and with equally satisfying results. One is simply to huddle. By huddling with another warm body, the temperature of each animal's surroundings is effectively raised (on the side of contact), thereby reducing the temperature gradient that drives heat loss, at least for a portion of the body. But "simply" huddling is greatly misleading, for huddling, first of all, implies acceptance of another's company in close quarters, and for a normally solitary vole this requires a significant change in social behavior—a serious attitude adjustment, if you will.

Social aggression in most animals is a programmed trait that serves a specific purpose and is not easily changed, yet several species of voles abandon their solitary ways in autumn for the energetic advantages of communal nesting. The relaxation of social aggression and territoriality is as least partly tied to cessation of reproductive activity and atrophy of sex organs, and this in turn is under hormonal control, influenced by changing day length and pineal melatonin levels (p. 8). But other factors may be at work, too. A high degree of interrelatedness (primarily between mothers and offspring), high population density (in which offspring are tolerated more), clumped food resources, and scarce availability of overwintering sites may all facilitate social aggregation in the fall. Conversely, low food availability and high metabolic demands (linked to very small body size) appear to preclude a changeover to communal nesting behavior in other small mammals, especially shrews.[1]

The advantage of several animals' building a communal nest and huddling through frosty nights is clear. Not only is there an energetic savings incurred through reduced surface area of exposure and increased insulative benefits of the nest itself, but whenever an animal ventures out, it always returns to a warm nest and therefore is not repeatedly giving up heat to a cold sink.[2] But this brings up another issue. These animals *are* active beneath the snow pack throughout winter and must forage periodically. Only now there are several individuals competing directly for resources in a subnivean world where food is in diminishing supply and their foraging range restricted. How do they manage to coexist without eventually showing signs of nutrient stress and aggression?

The answer to this dilemma, it appears, is for each nestmate to lose body mass before winter, thereby decreasing its total food requirement and minimizing the adverse effects of competition during communal nesting. As the nights of autumn grow longer, voles, instead of foraging more and gaining weight, actually undergo a systematic reduction in body mass. By the time the last leaf has fallen, most voles will weigh about 20% less than they did at the start of the season. So programmed is this response that an individual born in late summer and still actively gaining weight, even if not fully mature when autumn comes, will cease growth prematurely and slip into the same declining weight pattern as earlier born individuals.[3] Even captive voles given unlimited access to food show the same response. And there is nothing superficial about the weight loss. Most of it is associated with weight and volume reductions of internal organs, including the liver, adrenal gland, and reproductive tissues. In several (perhaps all) voles and shrews even the brain undergoes a significant mass reduction, up to 30% in some cases. This is accompanied by an actual reduction in skull size through resorption of bone tissue along the cranial sutures.[4] Only the heart is spared and, in some species at least, may actually undergo a weight increase in autumn.[5]

Decreasing day length alone appears to provide the stimulus for these changes, and the mechanism by which weight is lowered is simply through reduced food intake. Exposure to artificially low temperatures while an animal is being held under constant long-day conditions will not trigger the response, and no change in quality or quantity of food seems to reverse the process once started. The matter is again under hormonal control, and the critical day length, by what evidence is available, appears on the order of 12 to 13 hours.[6] This means that by the first part of September, voles throughout their range are already undergoing significant change.

What is the advantage of their losing weight instead of gaining it in autumn? The often quoted Bergmann's Rule of size tells us that in cold climates an animal should benefit from being larger because its increased mass relative to surface area confers greater cooling resistance. If this is the case, then our vole would be at a considerable disadvantage for its small size, and all the more so for losing mass in the fall. But it can also be argued that the larger animal has a greater total energy requirement than a smaller one and therefore is at a disadvantage in having to find more to eat at a time of year when food is often scarce. And therein perhaps lies the advantage to our vole. While in the close company of others it appears better off reducing its food requirements and minimizing competition with its nestmates in exchange for the marked benefits of communal nesting.[7] The energetic arguments of Bergmann's Rule are satisfied (the individual gives up some small size advantage in autumn, but the group gains thermal mass) and the voles are able to coexist by sharing resources. This is not the last word, however.

Because neither communal dwelling nor body size reduction carries any thermal benefits outside the nest, it is incumbent on the vole to spend as many hours in the nest as possible once cold weather arrives—which means minimizing time spent foraging. The latter may be accomplished either through fasting, at least during periods of greatest stress, and sub-sidizing nutritional requirements with fat reserves, or by storing food nearby. Since fattening was clearly not the evolutionary route taken by the vole (other considerations not withstanding, the accumulation of large subcutaneous fat deposits is rather cumbersome for an animal of this size), these animals have, for the most part, become dedicated food hoarders.

Voles will stockpile virtually any plant material manageable, includ-ing roots, bulbs, corms, stems, leaves, fruits, and nuts (though not often smaller seeds), and will also store fungi when available. Food is cached in larders—which minimizes recovery time compared to scatter hoard-ing—with rootstocks kept in shallow, underground chambers and above-ground plant parts stored in piles on the surface (in what seems an inordinate determination to store food items in a setting similar to that in which they were found, the bank vole of boreal forest regions will even hoard arboreal lichens in treetop caches, risking exposure during winter to retrieve them[8]). And voles acting in a community effort can stockpile impressive amounts of food for winter. The singing vole of northern Canada and Alaska constructs "haypiles" under overhanging rocks in much the same fashion that pikas do, sometimes containing up to a bushel of plant material.[9] Jerry Wolff, a mammalogist working in Fairbanks, Alaska, found at one of his study plots 36 quarts of fireweed and horsetail rhizomes put up by an estimated 6 to 10 taiga voles.[10] And one of the more notable accounts in the literature has a social group of a dozen or so Brandt's voles in Mongolia putting up in autumn 70 pounds of dried leaves, stems, and roots in subterranean chambers.[11]

Food hoarding among voles is often delayed until snow covers the ground, perhaps to evade predators or cache robbers. While evidence for such activity is usually found only after snowmelt in the spring, un-usual climatic circumstances recently enabled Jim and Audrey Bene-dict to document the initiation of undersnow caching behavior by montane voles at one of their study sites in the Colorado Rocky Mountains. The winter snowpack had already begun to form, com-pletely covering their site in a subalpine meadow by the end of Sep-tember, and, for all appearances, was in place for the season. However, unusual weather events in early October resulted in the complete loss of the snowpack, leaving the ground bare again after only 3 weeks of "winter." In that short period of time, montane voles had developed an extensive system of runways and root-foraging trenches, and had begun food hoarding activity at four separate larders. Already they had accumulated 887 tuberous roots from the pigmy bitterroot plant, an aggregate energy store of 395 kilocalories (kcal), which they simply heaped into piles beneath the snow.[12]

The significance of this energy investment can easily be translated into future "nest days" when the food stores might be utilized under more stressful conditions. Measurements of basal metabolism in field-caught red-backed voles indicate that individuals having undergone systematic weight reduction in autumn benefit from an approximate 30% savings in resting energy requirement under cold exposure. In specific terms this means that at an ambient temperature just under 0°F a winter-acclimatized vole would have to meet a daily need of about 24 kcal, compared to 33 kcal for a summer-weight vole.[13] At 32°F, however, a temperature more typical of a well-made, communally occupied nest beneath the snowpack, the vole would need fewer than 18 kcal to balance its heat loss—which means that the Benedict's montane voles had accumulated, in 3 weeks or less of hoarding, sufficient reserves to support one animal for about 22 more days with no additional foraging.

Even at this, the value of the voles' larder, in terms of nest time gained, is likely underestimated, for the voles have still other metabolic strategies for balancing the energy budget and reducing their foraging needs in the cold. While the voles are busy caching food in autumn, they are also quietly building up small but energy-packed brown fat reserves in critical parts of their body to fuel the metabolic furnace when the temperature plummets.[14] Brown fat is distinctly *un*like white fat—which is primarily an insulating layer and nutritional reserve—and serves only one purpose, to generate heat. If white fat can be thought of as an insulating blanket in some animals, then brown fat might be likened to an electric blanket, for when brown fat is metabolized, it yields heat—lots of it. And when the temperature drops low enough that curling, fluffing fur, or huddling is no longer sufficient to balance increasing heat loss to the animal's surroundings, the hypothalamus—the brain's temperature control center—signals the release of noradrenaline into the bloodstream from the adrenal gland, to activate brown fat metabolism. Heat is generated immediately, without shivering: enough, in fact, that subcutaneous temperatures measured over brown fat deposits can sometimes be observed to exceed body core temperature.

If that is not enough, there is only one involuntary course of action left. When, under the most threatening of circumstances, cold exposure is so great that none of the advantages so far encountered are enough to balance heat loss, then the animal begins to shiver. Shivering generates heat through work—through opposing muscle contractions that oxidize glucose rapidly—like a fire burning in muscle tissue. But also like fire, the rate at which heat can be produced in the muscle is limited by the supply of oxygen to the tissue. In order to get more heat out of shivering, then, the vole in autumn steps up production of muscle myoglobin, another oxygen-binding protein like hemoglobin, to help deliver the goods. This calls for improved cardiac output as well, possibly explaining the autumn increase in heart mass that has been observed in voles. Enhanced heat production through shivering is the last weapon in the vole's strategic arsenal against impending cold, but added to the others it provides an impressive backup. It is little wonder, then, that when Dale Feist at the University of Alaska first measured peak metabolic rates of cold-exposed voles live-trapped in autumn, he found a remarkable increase in their capacity to oxidize glucose and generate heat. Following a series of such measurements, Feist concluded that red-backed voles, after acclimitizing in the fall, are capable of matching heat loss to air at a temperature as low as −80°F for short periods of time.[15] It is an impressive ability indeed for such a small animal, but considering the lengths to which voles go in autumn to avoid such extremes (though evidence for communal nesting among red-backed voles is equivocal[16]), temperatures on that order become purely hypothetical for most.

.   .   .

Are the behavioral and metabolic changes displayed by voles typical of other nonhibernating mammals in the fall? The short answer is, "mostly, but not all." The majority of small mammals studied in temperate and northern latitudes employ some combination of these strategies to compensate for increased energy demands in the cold, though seasonal increases in pelage insulation, brown fat, and muscle myoglobin are probably the only ones common to all. Many species adopt communal nesting behavior, but lemmings and the small carnivores (the shrew *Cryptotis parva* and two other European species are noteworthy exceptions) do not. Autumn weight loss is known for a few other species, including the shrews that do not compensate for increased heat loss with huddling behavior, but has not been observed in the collared lemming.[17] Most species hoard food to one degree or another, including the carnivores, but there are notable exceptions like the muskrat, which only rarely has been known to cache plant material, even though it remains active and must forage under ice all winter.

There is, surprisingly, one strategy common among small mammals that voles do not employ, and that is the use of torpor—the short-term depression of body temperature—to reduce heat loss. By lowering body temperature 10 to 15 degrees, hence reducing the temperature gradient between the body core and its surroundings, an animal can cut its losses on a cold night by as much as 20%. Curiously, most of the rodents, as well as a number of birds, utilize some degree of controlled hypothermia to save energy, but no member of the mammalian subfamily Microtinae, the voles, has yet been shown to have this ability.

The range of overwintering preparations displayed by small mammals, and the crossovers in strategy—the communal nesting of one species of shrew, but not others; the autumn size reductions in some solitary animals, but not in social lemmings; the inability of voles to enter torpor while almost all other small mammals can—highlight an important aspect of environmental adaptation: In a world of seemingly infinite variation, there are no simple answers to the multitude of complexities and challenges that animals face. Natural selection may favor certain adaptive themes—behaviors that increase foraging efficiency or decrease energy loss, for example—but for every species there is a combination of traits that best suits its overall needs, a suite of adaptations that enables it to profitably exploit all the resources of its environment while at the same time avoiding predation and maximizing its reproductive success. No species seems to illustrate this point any better than the secretive—and successful—voles of northern lands.

Indian Summer

[W]e look in vain for any recognition of [Indian summer] in pages not more than half a century old. It seems to have departed from the land of the Puritans with the vanished forests, and doubtless these had much to do with its former prevalence.   WILLIAM HUBBARD, circa 1680[1]

&. Indian summer is a golden time. The sunlight seems softer, slightly paler than before, its warmth more coddling, more like the feel of wool than of summer's polyester. "The atmosphere holds the beams, and abstracts from them their white brilliance. They come slower with a drowsy light, which casts a less defined shadow of the still oaks. The yellow and brown leaves in the oaks, in the elms, and the beeches, in their turn affect the rays, and retouch them with their own hue."[2]

To most people, "Indian summer" is synonymous with the milder days of autumn. The term conjures up comfortable images of drying corn shocks and full, burnished orange pumpkins, of unseen pheasants honking from tall weeds beyond the field's edge, of soft, winey apples on the ground beneath forgotten trees. The air is fresh and crisp, but not wintry, with a slight pungence from the weathering of dried leaves. All the land is, as Richard Jefferies put it, "nutty, brown, hard, frosty, and sweet . . . where every step is bracing."[3] But these images, evocative though they may be, are only vignettes of autumn in general and not of Indian summer. Indian summer is—or was—something more specific.

The precise origin of the term "Indian summer" has been obscured in the few centuries since it was first coined, its meaning slowly changed, and there is even a possibility that the specific weather phenomenon given this name no longer exists. Use of the term traces back to the settlement of New England, where it appears to have had more of a historical connotation than reference to a particular set of climatic conditions. From early colonial records, nineteenth-century natural historian Charles Abbott produced the following account:

In the early periods of our history, when the Indian enemies lurked in the forests and burst out from their ambuscades on the planter, the first settlers enjoyed little security, except in the winter, when the severity of the season prevented the incursions of the savages. The coming of winter was hailed as the commencement of peace by the early inhabitants of the country; they sallied out from the little forts and block-houses, in which they had been hemmed up, with the joyful feelings of prisoners escaping from confinement, and busily gathered in their harvests. To our ancestors the snows of winter were more pleasant than the flowers of spring, as they brought the cessation of the horrors of war. But it often happened that the mild day of November afforded the red men another opportunity of visiting the settlements with those desolating blows, which burst like the lightening from the cloud, leaving the record of their effects in the blaze that followed the stroke. The activity of the red men during these periods gave, as is supposed, the name of "Indian summer" to those bright days, when autumn bestows its last parting favors."[4]

Though not referenced fully by Abbott, it appears that the source of this quotation is a 1677 document written by William Hubbard, a minister in Ipswich, Massachusetts, providing one of the earliest explanations of the term "Indian summer."[5] It is curious, however, that by the time of Abbott's writing, others among natural historians were already interpreting the term differently. John Burroughs, an Abbott contemporary, reflecting the tenor of two centuries' passing (and evidently without the benefit of history), expressed a much more genial view of its meaning:

This halcyon period of our autumn, will always in some way be associated with the Indian. It is red and yellow and dusky like him. The smoke of his campfire seems again in the air. The memory of him pervades the woods, His plumes and moccasins and blanket of skins form just the costume the season demands. It was doubtless his chosen period . . . if the red aborigine ever had his summer of fullness and contentment, it must have been at this season, and it fitly bears his name.[6]

Fifty years later we read from Edwin Way Teale[7] that early colonists attributed the characteristic haze of Indian summer, as well as its warmth, to aboriginal fires associated with the burning of prairie grasses—"hence the name of the season."

Somewhere in the transition from historical commentary to climatic event, the term took on rather specific meaning. The first clear delineation of its character from a natural historian came in the middle of the nineteenth century, when P. H. Gosse, a British naturalist who traveled widely in North America, described the phenomenon as follows:

It is observable that after all this short-lived splendour has passed away, and the trees have become leafless, in Canada and the Northern States, there always occur a few days of most lovely and balmy weather, which is called the Indian Summer. It is characterized by a peculiar haziness in the atmosphere, like a light smoke, by a brilliant sun, only slightly dimmed by this haze, and by a general absence of wind. It follows a short season of wintry weather, so as to be isolated in its character. One circumstance I have remarked with interest,—the resuscitation of insect life in abundance. Beautiful butterflies swarm around the leafless trees; and moths in multitudes flit among the weeds and bushes, while minuter forms hop merrily about the heaps of decaying leaves at the edges of the woods. It is a charming relaxation of the icy chains of winter.[8]

Confusion still persisted in early writings, though, and it wasn't long after Gosse's commentary that Charles Abbott complained, "the impressions of a dozen authors that I have collated, as to its time, place, and circumstance, are quite as hazy as the brief 'summer' itself."[9] Nonetheless, at least two common denominators emerged from the early literature. One is that Indian summer was repeatedly described as occurring specifically in November; that "warm, hazy, *dolce far niente* days in October or December are simply so much good luck, but not typical Indian Summer."[10] The other is that a widespread and prevalent haze was characteristic of the season. This was often ascribed, as Teale noted, to the burning of grasses or forest undergrowth by indigenous peoples, but a few naturalists of the time suspected that it might have been of entirely natural cause instead—causes as much biological in origin as atmospheric. Abbott summarized contemporary thought in a few words when he wrote in 1889:

[T]he character of this haze is an open question. It is said to be animal life so minute as to escape microscopial examination—hypothetical creatures that make up in numbers what they lack in size, and at one time shake the atmosphere and obscure the sun. By many it is thought to be of vegetable origin; and by a great many, in a pompous manner, it is said to be "haze, and any fool knows what that is."

Abbott was perplexed by the phenomenon, though, and readily admitted to his confusion. In fact, he could not resist a personal rejoinder to this quoted statement, adding "I glory in being one of the fools that do not know what haze is."[11] But even as Abbott pondered these matters, the frequency of true Indian summers in the northeastern United States was diminishing, for Abbott admitted to having experienced few himself.

In all likelihood, the haze *was* "of vegetable origin," and its initial prominence in the northeast, as well as its subsequent diminution, apparently was related to changing forest conditions in New England during the first two centuries of settlement. Land clearing had been steadily increasing since the British colonies were founded in the late seventeenth century, and by the early 1800s had progressed to the remotest corners of the northern Appalachian Mountains. What was once nearly continuous forest cover was by then mostly hill farms with small woodlots scattered about. Yet even as late as the 1800s in northern parts of New England, Indian summers were common enough that they, with their attendant haze, still attracted notice. Zadock Thompson, an early nineteenth-century Vermont educator and astute observer of natural history, paid particular attention and eventually provided exceptional insight into the probable cause of the haze:

> From the earliest settlement of this country there have been observed a number of days, both in spring and autumn, on which the atmosphere was heavily loaded with smoke. The smoke has generally been supposed to result wholly from extensive burnings in some unknown part of the country. [But] . . . the amount of smoke has not always been greatest in those years in which burnings were known to be the most extensive; and . . . moreover . . . the atmosphere was usually most loaded with smoke in those autumns and springs which succeeded warm and productive summers. These circumstances have led us to the opinion that the atmosphere may, by its solvent power, raise and support the minute particles of decaying leaves and plants, with no greater heat than is necessary to produce rapid decomposition. When, by the united action of the heat and moisture . . . the leaves are separated into minute particles, we suppose these particles may be taken up by the atmosphere, before they are entirely separated into their original elements, or permitted to form new compounds. This process goes on insensibly, until, by some atmospheric change, a condensation takes place, which renders the effluvia visible, with all the appearance and properties of smoke.[12]

Thus did Thompson, in 1853, outline in essence what we know today about complex aerosol chemistry in the atmosphere. Hydrogen and carbon atoms, boiling off the drying vegetation in vapor form, recombined in the presence of sunlight with other elements to form aerosols—particles slightly larger than molecules but small enough to remain suspended in the air. As these aerosols grew in size or became more abundant, they scattered light, creating the characteristic haze. Without the immense annual litter production of the eastern and northern deciduous forests, however, the occurrence of the haze in autumn gradually diminished.

As Thompson was also chronicling the demise of northern New England forests and wildlife, he made a connection between the two phenomena and went on to suggest why Indian summers had "gradually become more and more irregular and less strikingly marked in their character, until they have almost ceased to be noticed."

> When our ancestors arrived in this country, the whole continent was covered with one uninterrupted, luxuriant mantle of vegetation, and the amount of leaves and other vegetable productions, which were then exposed to spontaneous dissolution upon the surface of the ground, would be much greater than after the forests were cut down and the lands cultivated. It is very generally believed, that our winds are more variable, our weather more subject to sudden changes, our annual amount of snow less and our mean annual temperature higher than when the settlement of the country was commenced. And causes, which would produce these changes, would, we believe, be sufficient to destroy, in a great measure, the peculiar features of our Indian summers. The variableness of the winds, occasioned by cutting down large portions of the forests, would of itself be sufficient to scatter and precipitate those brooding oceans of smoke.[13]

It seems remarkable now that Hubbard noted the same declining frequency of hazy Indian summer days in New England settlements that had been cleared almost two centuries earlier, and suspected the same cause.

Deforestation of New England reached its apex in the mid 1800s when, with only about 20% of the original forest cover remaining, the westward expansion of railroads and introduction of coal into the cities finally took pressure off diminishing fuelwood supplies. At about the same time, the California Gold Rush and Civil War started to draw young men off the farm, and with the subsequent opening of more fertile lands to the west, New England's marginal agricultural economy slid downhill quickly. Widespread farm abandonment accelerated—and forest recovery began, first with a proliferation of white pines and then with the hardwoods. Might we expect, then, the eventual return of classic Indian summers as described by Thompson, Gosse, and others to areas where the regrowth of deciduous forests is substantial? Throughout the Appalachian Mountains region and as far west as Minnesota, the hardwood forests have been steadily recovering since the early 1900s. New England, in fact, has seen a return in forest cover almost to that of its early colonial days. But the new forest is not the same as the old, nor is our atmosphere the same as it once was. For better or for worse, we have changed our world in uncertain ways, and it remains to be seen whether we will ever experience true Indian summer with any regularity again.

The Fall of the Leaf

One leaf falls in the stillness of the air slowly, as if let down by a cord of gossamer gently, and not as a stone falls—fate delayed to the last.

RICHARD JEFFERIES, 1900[1]

› In the rustle of an autumn breeze, a maple leaf breaks free and floats lightly to the forest floor. It weighs only three-quarters of a gram, not quite three-hundredths of an ounce. By season's end, however, when in Jefferies's words "the leaf crop is gathered to the great barn of the earth," about 1½ tons of dry foliage will have been put up for every acre of maple wood. Small branches, loose bark, bud scales, and flower parts will have contributed perhaps another 20% to the year's total. Even in our temperate and northern coniferous forests, an equivalent amount of leaf litter may fall annually as older, less efficient evergreen foliage is discarded.[2] All this adds up to a lot of fodder for the grazers in the barn, but nature is not unprepared for the harvest. The leaf falling to the ground is already seeded with saprophytic fungi to start the decomposition process. Fate delayed to the last wastes no time.

The cast of decomposer organisms waiting to assimilate this organic bounty into their own being is impressive indeed. If one were to gather together every arthropod, snail, earthworm, nematode, slime mold, fungus, and bacterium (to name a few) in an acre of productive soil, their combined mass would likely top 5 tons per acre. Earthworms alone, in rich leaf compost, can weigh in at 900 pounds per acre. When slugs and snails are at their peak, they may exceed half a million per acre, with a total live weight approaching 400 pounds. Unseen nematodes, among the smallest of the soil grazers, number on the order of 8 to 24 billion in an acre of soil, their combined mass but small measure of their ability to process organic matter.[3] The real heavyweights of the decomposer organisms, however, are the fungi. Moselio Schaechter, a Tufts University mycologist, has made a rough estimate of their mass worldwide and figures conservatively that at just under half a ton per acre for all the vegetated land on earth, their total weight adds up to more than 2 tons for every human on the planet (the earthworms that tipped the scale at 900 pounds per acre in deciduous forests are absent from many vegetated areas, particularly in arid regions and in the far north).[4] In a Michigan hardwood forest, the spreading hyphal net formed by a single clone of the wood-rotting fungus *Armillaria bulbosa* has been traced beneath 36 acres of woodland, its thread-like body weighing an estimated 10 tons.[5]

It is the collective appetite of these organisms for organic matter that keeps forest litter from piling ever deeper, year after year. And no less diversity would suffice for the job, as it takes the cutting, grinding, and digesting activity of myriad specialists to disassemble all this material. Yet one group—the fungi—carries the bulk of responsibility for cleanup. Much of the mass of leaf litter is composed of cellulose and lignin, and for the soil animals (termites being the notable exception) these compounds are notoriously difficult to break down. Fungi, on the other hand, are able to exploit nearly every form of organic carbon available in the natural world, largely through their ability to secrete extracellular enzymes that break down compounds too large otherwise to be absorbed into their hyphal tissues. It should not be surprising, therefore, to find fungi growing virtually everywhere that organic material, living or dead, is found—including the surface of healthy green leaves.

The establishment of fungal communities on living leaves in the forest canopy is an intriguing story suggestive of an odd partnership with higher plants, in which early benefits to the plant are traded for later benefits to the fungi. From the time a tender new leaf pushes through its protective bud scales into a world of intricately tangled food webs, it is colonized by saprophytic fungi. The fungi lay in waiting, with three or four species of yeasts (unicellular fungi) often taking refuge in the leaf buds themselves (they have even been found within the shoot primordia of some deciduous trees and in the developing cones of pines).[6] Others arrive quickly in the form of wind- or waterborne spores. For the most part, these fungi pose no greater burden to the emerging leaf than that of Spanish moss festooning the branches of live oaks in the deep South. Unlike the decomposers of the forest floor, the early colonists are using the leaf primarily as a platform from which to scavenge organic debris, and they may actually confer several benefits to the new leaf. It is an unexpected partnership—for the plant a bit like marrying the undertaker—but it seems to work for both.

One advantage to the plant of keeping company with fungi derives from a potent insecticidal property of the latter. On the leaves of many plants, these fungi have been found to secrete strong alkaloids that interfere with insect larval development—even killing the larvae outright—with the result that fewer leaf-chomping caterpillars feed their way into adulthood (and fewer fungi end up in the salad).[7] There is nothing altruistic about it, of course; the fungus is merely looking out for itself, and in the process reduces herbivory to the plant. In a related defense against potential competitors, these same fungi often display antagonistic effects on the growth of certain other microbes, which can reduce subsequent colonization of the leaf by potential parasites. Additional unknown advantages may be conferred, too. Grasses, for example, that harbor fungi in their intercellular spaces (fungi that move into leaf interstices via open stomata) often show increased vigor and greater productivity.[8] The fungi, for their part, gain access to organic particles on the leaf surface—energy sources like pollen, spores, or other wind-blown debris—as well as nutritious exudates that leak from the epidermal cells of the leaf.

What is good for some fungi is probably good for many others, and in nature, opportunity seldom goes unexploited for long. There are an estimated 1½ million fungal species in the world, many of which are occupational generalists and nearly all of which are adapted for wide dispersal. With billions of fungal spores in the air, the resident population on the leaf is quickly infiltrated by newcomers. The most successful of the early invaders appear to be members of the class Deuteromycetes (formerly known as "fungi imperfecti"), which are asexually reproducing fungi. Initial colonization is often dominated by *Cladosporium herbarum*, a species of notorious tolerance to extremes that grows in velvety green, branching chains, and thus is not likely to be noticed on the healthy leaf (one strain of *C. herbarum* attacks meat in cold storage and can thrive at below-freezing temperatures!).[9] *Cladosporium* may be quickly joined by other "weedy" microfungi, including species of *Penicillium* and *Fusarium*, the latter better known for its several parasitic strains that attack vegetable crops.[10]

Life on the surface of a leaf is not easy, though, and for many of the recent arrivals the stay is short. Viewed at the submicroscopic level, the epidermis of the leaf appears no more hospitable than the rugged mesa and canyon country of the arid southwest, to which it bears remarkable structural resemblance. What appears a smooth, flat leaf surface to the eye is instead a complex landscape of erosion features worn into the leaf cuticle, its outer waxy coating.[11] Lodged within the hidden canyons of the leaf are many of the organic resources needed to sustain fungal growth, yet relative to the topographic features of the leaf, the newly arrived spore sits exposed like a giant boulder on a desert landscape. Of all the spores that become attached to the leaf surface, surprisingly few are able to survive there. Competition and restricted growth often result from nutritional shortages, especially on the younger leaves. Midsummer brings frequent drought to a leaf surface that dries quickly, and thunderstorms can wash spores out off the leaf. The fungal colonists secrete sticky mucilaginous sheaths that help them hang on, but it's a tenuous existence; some populations die, others produce resting structures and weather the drought, recent arrivals try anew. It takes a hardy pioneering type to succeed under these circumstances.

By late summer, however, the rain of spores from the atmosphere becomes increasingly heavy. Through the shear force of numbers, new recruits on the leaf surface begin to outnumber losses. The same sticky mucilage that holds one fungus to the leaf also attaches other spores. Climate conditions tend to improve as well, with the cooler days resulting in reduced evaporation and the nights often bringing dew formation. These factors, along with the gradually increasing pollen deposition on the leaf surface—a ready source of energy for saprophytic fungi—provide all the resources needed for the establishment of a new community.[12] Many of the early colonists, aggressive species capable of growing fast under rigorous conditions, have by now formed resting spores and relinquished the site to superior competitors. As the aging leaf secretes more exudates, other species thrive. Succession is underway. The fungal community that once helped protect the leaf against predation has changed in its makeup, with the new dominants taking on a larger task. The return of nutrients to the soil has slowly but inexorably begun, high up in the forest canopy.

On the forest floor the environment of the leaf changes dramatically. No longer does it flutter in a drying atmosphere of sun and wind, but now it lies shingled with others to form an imperfect thatch over the ground, letting water percolate through but trapping moisture beneath in the dimly lit world of the organic litter layer. Some species of fungi whose presence on the leaf in the canopy was barely discernable may prosper now in the more humid environment of the forest floor, but many who rode the leaf to the ground, like messengers of fate, have delivered their promise and will take little part in decomposition of the litter. Though these fungi may persist well into the next year, most are poorly adapted for continued growth in their new environment, and relatively ineffective in breaking down leaf litter.[13] They will simply carry out their life cycles, largely unnoticed, maturing, fruiting, and spreading the spores of another generation.

The task of putting away the annual windfall of chemical energy and of returning nutrients to the soil for the next cycle belongs now mostly to the litter fungi. A few of the secondary saprophytes that colonized the leaf shortly before senescence will reach their full development on the ground, to be joined the following summer by still other litter inhabiting species. The latter will include a number of widely distributed fungi in the class Basidiomycetes, which form many of the familiar gilled mushrooms such as *Clitocybe, Collybia, Marasmus,* and *Mycena* (most have no common names). These are the heavy hitters, equipped with all the enzymes necessary to degrade lignin, disassemble the complex polysaccharides of the cell wall, and detoxify strong phenolic defense compounds in the leaf. As a group, these are perhaps the most active decomposers of leaf litter, commonly making up the bulk of the fungal biomass on the forest floor.[14] Basidiomycetes do not work alone, however. The rate at which fungi digest material is strongly dependent on the presence of soil microfauna, whose feeding activity breaks leaf material into smaller particles, compacting the litter, entrapping more moisture, and exposing more surface area to the enzymatic action of the fungi. The fungi, in turn, become an important source of digestible protein and carbohydrates for a myriad of soil insects who feed upon the hyphae permeating the litter,[15] ultimately releasing through their own waste products, nutrients temporarily sequestered by the fungi.

All of this activity—the grazing of earthworms and arthropods on leaf litter, the disassembling of complex carbohydrates by fungi, the breakdown of plant protein and amino acids by bacteria—is regulated to a large extent by physical aspects of the soil environment, especially its acidity, moisture content, and temperature. The former may determine what species of decomposers dominate the soil community, by their relative tolerances to low or high pH, but the latter two usually determine how quickly the annual leaf fall will be processed and its nutrients made available again to plants. Where moisture is not limiting, where neither drought nor waterlogging inhibits microbial respiration, and where temperatures are not so cold as to depress chemical reaction rates severely, the complete decomposition of organic material may occur within a matter of a few years. In the mixed deciduous forests of north-temperate regions the forest floor normally contains about a 10-year backlog of leaf litter in varying degrees of decay, with the soil organisms in some semblance of equilibrium, keeping pace, more or less, with each year's new additions. In the Pacific Northwest, however, where moisture is often severely limiting during the summer, a 50-year backlog of organic material normally exists. Here, the majority of decomposer activity occurs in the winter months beneath snow cover where soils are cold, but unfrozen, and more moisture is available. In the northern coniferous forests of Canada, Alaska, and Eurasia, where colder and often water-logged soils severely inhibit decomposition processes, microbes may never keep up with the annual litter fall, and the organic forest floor simply continues to build in thickness. A residence time of more than 350 years has been estimated for conifer litter in some northern forests.[16] Even at that, however, the amount of nutrients released in forested eco-systems annually through litter decomposition far outweighs any other source of input, such as from rainfall or rock weathering. In mixed deciduous-conifer forests of north-temperate latitudes, slightly more than 100 pounds per acre (total) of nitrogen, phosphorus, potassium, calcium, and magnesium are returned to the soil each year[17]—enough to satisfy approximately 75 to 85% of the trees' annual requirements for these essential elements.[18]

As plant litter is reduced to fine organic humus, microscopic fungi again assume importance, breaking down partially disintegrated lignin into its molecular components. But the work of the Basidiomycetes is not yet finished. It was they who first fractured the highly resistant lignin into its basic building blocks, and it is primarily they who will convey the final, mineralized chemical elements of the decomposed material back to the living plant. In one of nature's most remarkable partnerships, a great many Basidiomycetes form mycorrhizal (literally, "fungus root") associations in which they take up residence in plant roots for a ready supply of carbohydrates, but in so doing garner nutrients for the plant via their extensive hyphae, and do it far more efficiently than roots alone. Together, decomposer and mycorrhizal Basidiomycetes close the loop in the nutrient cycling process.[19]

Consider what a vast crop is thus annually shed upon the earth. This, more than any mere grain or seed, is the great harvest of the year. This annual decay and death, this dying by inches, before the whole tree at last lies down and turns to soil . . . I am more interested in it than in the English grass alone or in the corn. It prepares the virgin mould for future cornfields on which the earth fattens.

I pick up a white-oak leaf, dry and stiff, but yet mingled red and green, October-like, whose pulpy part some insect has eaten, beneath, exposing the delicate network of its veins. It is very beautiful held up to the light; such work as only an insect eye could perform. Yet, perchance, to the vegetable kingdom, such a revelation of ribs is as repulsive as the skeleton in the animal kingdom. In each case, it is some little gourmand working for another end, that reveals the wonders of nature.

—HENRY DAVID THOREAU, 1853, 1855[20]

With much of the work of fungi going on in the deeper litter layers, most of us rarely notice their presence, until mushrooms suddenly poke up through the moist duff. And often the mushrooms appear in autumn, as if they were quietly ushered in to help harvest the new leaf crop. But the falling of leaves and the increasing abundance of mushrooms in autumn woods are in part only coincidence, more a trick of the weather than a response to the bonanza of fresh organic matter.

Fungi, as we now know, are found almost everywhere and may persist out of sight indefinitely, as long as there is a source of organic energy to exploit. But when it comes time to reproduce, most fungi (excepting the yeasts and molds) betray their cover with a conspicuous and often brightly colored fruiting body, quite analogous to the fruits of flowering plants. The mushroom is a spore-bearing reproductive structure that can form whenever one wandering hyphal strand meets and pairs with another of compatible mating type. Pairing in this case involves fusing or growing together as in a graft of two branches or roots, since fungi have no distinguishable sexual parts. And the term "mating type" is substituted here for "gender" (though the latter is also acceptable) because our usual concept of gender—the simple notion of maleness and femaleness—is easily confused by the extraordinary number of mating types possible in a single species of fungus. Moselio Schaechter, the same mycologist who gave us the weight of all earth's fungi (p. 128), noted that the split-gill fungus, *Schizophyllum commune*, has enough genders to permit 21,000 different pairings, every one between hyphal strands of a nonidentical mating type![21]

The development of a mushroom, then, must await the chance meeting of two such hyphae in their random exploration of the litter layer. The odds of a successful pairing, however, are better than they might at first appear. With such a large number of mating combinations possible, a hyphal strand is far more likely to run into a compatible gender than one that is identical to itself. But even after pairing, the fruiting body does not develop until it receives the appropriate environmental cue— which most often is simply an abundance of moisture. The importance of moisture is a practical matter, based more on hydraulic needs than on support of cell respiration, for the seemingly overnight growth of mushrooms is primarily due to rapid water absorption and expansion of cells in an elongate direction, rather than a result of cell division and proliferation. And so common is this requirement that moisture sufficient to stimulate mushroom growth in one species often spurs other species in the neighborhood to fruit as well.[22]

The need for abundant moisture to reproduce also helps explain why in many cases mushrooms seem to appear in the fall of the year. Throughout much of temperate North America, autumn brings more frequent rainfall than occurs in summer. All across the northern states, from the Appalachians to the Pacific Northwest, the number of days with recorded precipitation increases steadily between the first of August and the end of November.[23] Along with the increase in cloud cover, greatly reduced evaporation and a reduction in water use by senescing plants adds considerably to the amount of water available in the soil and litter layer.

So the fungi, blanketed in organic riches and refreshed with abundant moisture, thrive and reproduce in the crisp air of autumn. And within days of fruiting they release their spores in powdery wisps to drift over the land, infusing the ranks of the silent decomposers with new strength, driving nutrient cycles ever forward to the benefit of every higher organism on earth that requires mineral elements for its own growth and development. Mushrooms are a symbol of renewal and a symbol of the season. *They* are the shining fruits of the great autumn leaf harvest.

Repose

Somewhere in the wood was the ghost of Proserpine returned to see how her children were faring—under the leaves were the seeds that would bring forth bloom and beauty and fragrance in the spring; deep in the earth lay the cocoons and shells whence would arise the happy throngs of summery moths and butterflies. For this is the purpose of autumn: rest and quietude for those who have labored throughout the summer to insure life for their kind. So now in the autumn my hope is as firm as the oak. Every leaf that falls is pushed from its hold by a bud awaiting the mystic order to unfold itself in spring.

HENRY WILLIAMSON, 1922[1]

The leaves are down now; the brilliant colors of fall slowly dissolve into the humus of the forest floor. The migrants have departed; the hibernators have lapsed into a deep coma. Those left behind to cope seem quiet, secretive, waiting, as autumn gradually hardens into winter. Only the brief flurry of a passing flock of chickadees and kinglets breaks the air of anticipation.

But the season does not surrender easily. Just as summer pushes insistently into autumn with the late flowering of goldenrod, asters, and witch hazel, so, too, will the beech and oaks stubbornly hold their leaves into winter, denying their inevitable loss. Wild apples will hang on bare branches until after the snows have arrived, as if in sympathy with winter browsers, and the last seeds of rigid brown flower stalks will not part until icy winds shake them stiffly over crusted snow.

For the apparent disappearance of many plants and animals, autumn is often seen as an end. But the seasons are part of a continuum, a revolving process of birth, death, and renewal—and if such could be said to have any beginning or end, then fall could just as well be viewed as a beginning. Like the counterweight on a flywheel, autumn provides the inertia to turn the annual cycle over. The seeds of another season have already been planted—sown on the wind and the wings of birds and the coats of animals to find new life in new places. Another generation is already awakening in the wombs of the great mammals. And in all the hidden sanctuaries of autumn—in the crevices of dormant trees, in the cold safety of piled leaves and decaying logs, in the sediments of stream and pond bottom—myriads of insect larvae are beginning their incredible metamorphic journey into spring and adulthood. Energy is flowing and nutrients are circulating. These are the processes by which nature's bounty is reinvested in a burst of new growth, reproduction, and dispersal, to arrive at yet another autumn and another season of change.

Notes

Preface

1. J. Kieran, *Footnotes on Nature* (Garden City, N.Y.: Doubleday, 1947).
2. C. Weygandt, *The White Hills: Mountain New Hampshire, Winnepesaukee to Washington* (New York: Henry Holt, 1934).
3. J. Burroughs, *Winter Sunshine* (Boston: Houghton, Mifflin, 1894), iiv.
4. J. Burroughs, *Under the Maples* (Boston: Houghton, Mifflin, 1924).

Keepers of Time

1. J. Burroughs, *Under the Maples* (New York: Wm. H. Wise and Co., 1924).
2. H. D. Thoreau, *Walden or Life in the Woods* (Boston: Houghton Mifflin, 1906), 327, 333.
3. Burroughs, *Under the Maples*.
4. G. C. Brainard, J. R. Gaddy, F. M. Barker, J. P. Hanifin, and M. D. Rollag, "Mechanisms in the Eye that Mediate the Biological and Therapeutic Effects of Light in Humans," in L. Wetterberg, ed., *Light and Biological Rhythms in Man* (Oxford: Pergamon Press, 1993), 29–54.
5. Ibid., p. 42.
6. B. K. Follett, "Photoperiodic Physiology in Animals," in J. Brady, ed., *Biological Timekeeping* (Cambridge: Cambridge University Press, 1982), 83–99.
7. S. Binkley, *The Pineal: Endocrine and Nonendocrine Function* (Englewood Cliffs, N.J.: Prentice Hall, 1988), 262–282. For a complete synthesis of endocrine control of animal behavior see R. J. Nelson, *Behavioral Endocrinology* (Sunderland, Mass.: Sinauer Associates, 1995).
8. R. J. Reiter, "The Melatonin Rhythm: Both a Clock and a Calendar," *Experientia* 49 (1993): 654–665.
9. Ibid.
10. Ibid., and Binkley, "The Pineal," 271–278.
11. I. Ting, *Plant Physiology* (Reading, Mass.: Addison Wesley, 1982), 548.
12. Ibid.
13. F. Salisbury and C. Ross, *Plant Physiology* (Belmont, Calif.: Wadsworth, 1992), 446.
14. B. Thomas and D. Vince-Prue, *Photoperiodism in Plants* (New York: Academic Press, 1997), 106–109.
15. D. Dennis and D. Turpin, eds., *Plant Physiology, Biochemistry, and Molecular Biology* (New York: Longman Scientific and Technical and John Wiley and Sons, 1990), 40.
16. K.-A. Stokkan, N. J. C. Tyler, and R. J. Reiter, "The Pineal Gland Signals Autumn

to Reindeer *(Rangifer tarandus tarandus)* Exposed to the Continuous Daylight of the Arctic Summer," *Canadian Journal of Zoology* 72 (1994): 904–909.

17. M. McCulloch Williams, *Field-Farings* (New York: Harper and Brothers, 1892), 154–155, 159–160.

Turning

1. L. D. Nooden and A. C. Leopold, *Senescence and Aging in Plants* (New York: Academic Press, Inc., 1988): 6–13.
2. F. Stuart Chapin III and R. A. Kedrowski, "Seasonal Changes in Nitrogen and Phosphorus Fractions and Autumn Retranslocation in Evergreen and Deciduous Taiga Trees," *Ecology* 64 (1983): 376–391.
3. Ibid.
4. J. Burroughs, *Under the Maples* (New York: Wm. H. Wise & Co., 1924).
5. P. Nobel, *Biophysical Plant Physiology and Ecology* (San Francisco: W. H. Freeman and Co., 1983), 252–254.
6. F. Salisbury and C. Ross, *Plant Physiology* (Belmont, Calif.: Wadsworth, 1992), 324–325.
7. Ibid.
8. C. Abbott, *Days Out of Doors* (New York: D. Appleton and Co., 1889), 256.
9. Ibid.
10. Ibid.
11. R. Jefferies, *Field and Hedgerow* (London: Longmans, Green and Co., 1900), 157.
12. M. McCulloch Williams, *Field-Farings* (New York: Harper and Brothers, 1892), 160–161.
13. R. L. Stevenson, *Essays of Travel* (London: Chatto and Windus, 1905), 109.
14. R. H. Waring and W. H. Schleshinger, *Forest Ecosystems* (Orlando, Fla.: Academic Press, 1985), 176.
15. K. Esau, *Plant Anatomy* (New York: John Wiley and Sons, 1965), 470–472.
16. Nooden and Leopold, *Senescence and Aging in Plants*, 13, 330–331.
17. Esau, *Plant Anatomy*, 471.
18. Ibid., 472.
19. P. Host, "Effect of Light on the Moults and Sequences of Plumage in the Willow Ptarmigan," *Auk* 59 (1942): 388–403.
20. L. L. Badura and B. D. Goldman, "Prolactin-Dependent Seasonal Changes in Pelage: Role of the Pineal Gland and Dopamine," *Journal of Experimental Zoology* 261 (1992): 27–33.
21. C. F. Bassett and L. M. Llewellyn, "The Molting and Fur Growth Pattern in the Adult Mink," *American Midland Naturalist* 42 (1949): 751–756.
22. J. H. Severaid, "Pelage Changes in the Snowshoe Hare," *Journal of Mammalogy* 26 (1945): 41–63.
23. J. K. Terres, *The Audubon Society Encyclopedia of North American Birds* (New York: Alfred A. Knopf, 1980), 616.
24. H. D. Thoreau, *Walden or Life in the Woods* (Boston: Houghton Mifflin, 1906).
25. R. C. Gruys, "Autumn and Winter Movements and Sexual Segregation of Willow Ptarmigan," *Arctic* 46 (1993): 228–239.

26. S. Hadwen, "Color Changes in *Lepus americanus* and Other Animals," *Canadian Journal of Research* 1 (1929): 189–200.
27. J. K. Terres, *The Audubon Encyclopedia of North American Birds* (New York: Alfred A. Knopf, 1980).
28. L. Martinet, D. Allain, and C. Weimer, "Role of Prolactin in the Photoperiodic Control of Moulting in the Mink (*Mustela vison*)," *Journal of Endocrinology* 103 (1984): 9–15.
29. For a more detailed discussion of seasonal insulative adjustment and its relation to animal energetics in winter, see P. Marchand, *Life in the Cold* (Hanover, N.H.: University Press of New England, 1996), chapter 4 and the numerous references cited therein.

A Touch of Frost

1. J. P. Palta and P. H. Li, "Cell Membrane Properties in Relation to Freezing Injury," in P. H. Li and A. Sakai, eds., *Plant Cold Hardiness and Freezing Stress: Mechanisms and Crop Implications* (New York: Academic Press, 1978).
2. A. J. Hirsh, R. J. Williams, and H. T. Meryman, "A Novel Method of Natural Cryoprotection," *Plant Physiology* 79 (1985): 41–56.
3. For a more detailed discussion of freezing processes in plants and insects, see P. Marchand, *Life in the Cold* (Hanover, N.H.: University Press of New England, 1996), 43–49, 127–129, as well as the numerous references contained therein.
4. Ibid.
5. K. B. Storey and J. M. Storey, "Freeze Tolerance and Freeze Avoidance in Ectotherms" in C. H. Wang, ed., *Comparative and Environmental Physiology 4: Animal Adaptation to the Cold* (Berlin: Springer-Verlag, 1989), 51–82.
6. J. Rickards, J. J. Kelleher, and K. B. Storey, "Strategies of Freeze Avoidance in Larvae of the Goldenrod Gall Moth, *Epiblema scudderiana*: Winter Profiles of a Natural Population," *Journal of Insect Physiology* 33 (1987): 443–450.
7. Storey and Storey, "Freeze Tolerance and Freeze Avoidance," 51–82, and R. A. Ring, "The Physiology and Biochemistry of Cold Tolerance in Arctic Insects," *Journal of Thermal Biology* 6 (1981): 219–229.
8. J. G. Dunman, J. P. Morris, and F. J. Castellino, "Purification and Composition of an Ice Nucleating Protein from Queens of the Hornet, *Vespula maculata*," *Journal of Comparative Physiology* 154 (1984): 79–83.
9. Storey and Storey, "Freeze Tolerance and Freeze Avoidance," 51–82.
10. J. G. Dunman, "Change in Overwintering Mechanism of the Cucujid Beetle, *Cucujus clavipes*," *Journal of Insect Physiology* 30 (1984): 235–239.

Reflections on the Pond

1. J. R. Hazel, "Cold-Adaptation in Ectotherms: Regulation of Membrane Function and Cellular Metabolism," in C. H. Wang, ed., *Advances in Comparative and Environmental Physiology 4: Animal Adaptation to Cold* (Berlin: Springer-Verlag, 1989), 1–50.
2. K. Schmidt-Nielson, *Animal Physiology: Adaptation and Environment* (New York: Cambridge University Press, 1997), 184–186.

3. M. W. Oswood, L. K. Miller, and J. G. Irons III, "Overwintering of Freshwater Benthic Macroinvertebrates," in R. E. Lee, Jr., and D. L. Denlinger, eds., *Insects at Low Temperature* (New York: Chapman and Hall, 1991), 360–365.

4. K. B. Storey and J. M. Storey, "Triggering of Cryoprotectant Synthesis by the Initiation of Ice Nucleation in the Freeze Tolerant Frog, *Rana sylvatica*," *Journal of Comparative Physiology* 156(1985): 191–195.

5. J. R. Layne, Jr., R. E. Lee, Jr., and T. L. Heil, "Freezing-Induced Changes in the Heart Rate of Wood Frogs (*Rana sylvatica*)," *American Journal of Physiology* 257 (1989): R1046–R1049.

6. Ibid.

7. J. M. Storey and K. B. Storey, "Out Cold: The Winter Life of Painted Turtles," *Natural History* 1 (1992): 23–25.

8. Ibid., 25.

Down the Long Wind

1. W. Beebe, *The Log of the Sun* (Garden City, N.Y.: Garden City Publishing Co., 1906), 271.

2. P. Berthold, *Control of Bird Migration* (London: Chapman & Hall, 1996), 3–4.

3. Ibid., 8.

4. Ibid., 9.

5. J. C. Welty, *The Life of Birds* (Philadelphia: W. B. Saunders, 1962), 456.

6. Berthold, *Control of Bird Migration*, 7.

7. Owen & Black, 1991, cited in Berthold, *Control of Bird Migration*, 7.

8. A. Lindstrom, "Finch Flock Size and Risk of Hawk Predation at a Migratory Stopover Site," *Auk* 106 (1989): 225–232.

9. J. A. Thomson, *The Biology of Birds* (New York: Macmillan Co., 1923), 168. The source of the last three lines in quotes was not given, unless they are by Thomson himself.

10. M. Ramenofsky, "Fat Storage and Fat Metabolism in Relation to Migration," in E. Gwinner, ed., *Bird Migration: Physiology and Etophysiology* (Berlin: Springer Verlag, 1990), 220.

11. Ibid., 215.

12. W. H. Karasov, "In the Belly of the Bird," *Natural History* November (1993): 32–36.

13. P. Berthold, *Bird Migration* (Oxford: Oxford University Press, 1993), 102–105.

14. Ramenofsky, "Fat Storage and Fat Metabolism," 223.

15. H. Winkler and B. Leisler, "On the Ecomorphology of Migrants," *Ibis*, 134 Suppl. 1 (1992), 21–28.

16. Welty, *The Life of Birds*, 457.

17. F. M. Chapman, "The Post-Glacial History of *Zonotrichia capensis*," *Bulletin of American Museum of Natural History*, 77 (1940): 381–438.

18. Winkler and Leisler, "On the Ecomorphology of Migrants," 23.

19. J. K. Terres, *Audubon Society Encyclopedia of North American Birds* (New York: Alfred A. Knopf, 1980), 616.

20. H. Williamson, *The Lone Swallows* (London: W. Collins Songs & Co., 1922), 166–171.

21. Peter Marchand, previously unpublished.
22. W. J. Richardson, "Timing of Bird Migration in Relation to Weather: Updated Review," in E. Gwinner, ed., *Bird Migration: Physiology and Ecophysiology* (Berlin: Springer-Verlag, 1990), 84.
23. Berthold, *Control of Bird Migration*, 188.
24. Richardson, "Timing of Bird Migration," 88.
25. Berthold, *Control of Bird Migration*, 194.
26. Richardson, "Timing of Bird Migration," 95.
27. Ibid., 93.
28. Ibid., 90, citing various references.
29. Berthold, *Control of Bird Migration*, 35–36.
30. Alerstam, 1990, and Hummel & Benkenberg, 1989, in Berthold, *Control of Bird Migration*, 157.
31. Beebe, *Log of the Sun*, 269.
32. R. Mazzeo, "Homing in the Max Shearwater," *Auk* 70 (1953): 200–201.
33. L. R. Mewaldt, "California Sparrows Return from Displacement to Maryland," *Science* 146 (1964): 941–942.
34. Perdeck's experiments are summarized in G. V. T. Matthews, *Bird Navigation* (Cambridge: Cambridge University Press, 1968).
35. Kramer's work, published in German, is summarized in Berthold, *Bird Migration*, 147.
36. T. H. Waterman, *Animal Navigation* (New York: W. H. Freeman, 1989), 112.
37. K. P. Able and M. A. Able, "Calibration of the Magnetic Compass of a Migratory Bird by Celestial Rotation," *Nature* 347 (1990): 378–389, and "Daytime Calibration of Magnetic Orientation in a Migratory Bird Requires a View of Skylight Polarization," *Nature* 364 (1993), 523–525.
38. Waterman, *Animal Navigation*, 110.
39. K. P. Able, "Magnetic Orientation and Magnetoreception in Birds," *Progress in Neurobiology* 42 (1994): 449–473.
40. Terres, *Encyclopedia of North American Birds*, 607.
41. Waterman, *Animal Navigation*, 164.
42. Ibid., 166.
43. See various studies summarized by Berthold, *Control of Bird Migration*, 256.
44. H. D. Thoreau, 1857, appearing in H. G. O. Blake, ed., *Autumn: From the Journal of Henry D. Thoreau* (Boston: Houghton, Mifflin and Company, 1892), 231.

The Improbable Flight of Insects

1. K. S. Hagen, "Biology and Ecology of Predaceous *Coccinellidae*," *Annual Review of Entomology* 7 (1962): 289–326.
2. R. E. Lee, Jr., "Aggregation of Lady Beetles on the Shores of Lakes," *American Midland Naturalist* 104 (1980): 295–304.
3. Hagen, "Biology and Ecology of Predaceous *Coccinellidae*."
4. J. R. Riley et al. (9 authors), "The Long-Distance Migration of *Nilaparvata lugens* (Stal) (*Delphacidae*) in China: Radar Observations of Mass Return Flight in the Autumn," *Ecological Entomology* 16 (1991): 471–489.
5. S. J. Johnson, "Insect Migration in North America: Synoptic-Scale Transport in a

Highly Seasonal Environment," in V. A. Drake and A. G. Gatehouse, eds., *Insect Migration: Tracking Resources Through Space and Time* (Cambridge: Cambridge University Press, 1995), 55–57.

6. P. Street, *Animal Migration and Navigation* (New York: Charles Scribner's Sons, 1976), 75–77.
7. S. Benvenuti, P. Dall'Antonia, and P. Ioale, "Directional Preferences in the Autumn Migration of the Red Admiral (*Vanessa atalanta*)," *Ethology* 102 (1996): 177–186.
8. T. Larsen, "Butterfly Mass Transit," *Natural History* June (1993), 31–38.

Walking the Whole Way South

1. K. Schmidt-Nielson, *Animal Physiology: Adaptation and Environment* (Cambridge: Cambridge University Press, 1983), 211–212.
2. D. I. Beck, "Black Bears of West-Central Colorado," *Colorado Division of Wildlife Technical Pub.* 39 (1991), 86 pp.
3. For a review of the literature on cold tolerance and winter adaptation in Caribou, see P. Marchand, *Life in the Cold* (Hanover, N.H.: University Press of New England, 1996), 207–211.
4. R. D. Boertje, "An Energy Model for Adult Female Caribou of the Denali Herd, Alaska," *Journal of Range Management* 38 (1985): 468–73.
5. J. P. Kelsall, *The Migratory Barren-Ground Caribou of Canada* (Ottawa: Roger Duhamel, F.R.S.C., 1968): 129.
6. Ibid., 171.
7. Ibid., 171.
8. H. Stuck, *Ten Thousand Miles With a Dog Sled* (Prescott, Ariz.: Wolfe Publishing Co., 1988), originally published in 1914.

Deep Rut

1. M. P. Skinner, *The Yellowstone Nature Book* (Chicago: A. C. McClurg & Co., 1924), 186–187.
2. A. deVos, P. Brokx, and V. Geist, "A Review of Social Behavior of the North American Cervids During the Reproductive Period," *American Midland Naturalist* 77 (1967): 398.
3. Ibid., 399.
4. P. C. Lent, "Rutting Behavior in a Barren-Ground Caribou Population," *Animal Behaviour* 13 (1965): 259–265.
5. V. Geist, "Adaptive Behavioral Strategies," in J. W. Thomas and D. E. Toweill, eds., *Elk of North America: Ecology and Management* (Harrisburg, Pa.: Stackpole Books, 1982), 233.
6. Ibid., 230.
7. D. Petersen, *Among the Elk* (Flagstaff, Ariz.: Northland, 1988), 29.
8. Lent, "Rutting Behavior," 260–261.
9. V. Geist, " Behavior: Adaptive Strategies in Mule Deer," in O. C. Wallmo, ed., *Mule and Black-Tailed Deer of North America* (Lincoln, Neb.: University of Nebraska Press, 1981), 196.
10. Geist, "Adaptive Behavioral Strategies," 260.

A Dangerous Chill

1. J. Burroughs, *A Year in the Fields* (Boston: Houghton, Mifflin and Co., 1897), 176.
2. R. J. Mackie, K. L. Hamlin and D. F. Pac, "Mule Deer," in J. A. Chapman and G. A. Feldhamer, eds., *Wild Mammals of North America* (Baltimore, Md.: Johns Hopkins University Press, 1982), 862.
3. E. H. McEwan and P. E. Whitehead, "Seasonal Changes in the Energy and Nitrogen Intake in Reindeer and Caribou," *Canadian Journal of Zoology* 48 (1970): 905–913.
4. For a comprehensive review of the literature see P. J. Marchand, *Life in the Cold* (Hanover, N.H.: University Press of New England, 1996), 190–211.
5. B. Barnes, C. Omtzigt, and S. Daan, "Hibernators Periodically Arouse in Order to Sleep," in C. Carey, G. Florant, B. Wunder, and B. Horwitz, eds., *Life in the Cold: Ecological, Physiological and Molecular Mechanisms* (Boulder, Colo.: Westview Press, 1993), 164.
6. A. R. French, "The Patterns of Mammalian Hibernation," *American Scientist* 76 (1988): 569–575.
7. B. Barnes and D. Ritter, "Patterns of Body Temperature Change in Hibernating Ground Squirrels: Males Store Food," in Carey et al., *Life in the Cold*, 119–130.
8. French, "Patterns of Mammalian Hibernation," 574.
9. G. R. Michener, "Sexual Differences in Hibernaculum Contents of Richardson's Ground Squirrels: Males Store Food," in Carey et al., *Life in the Cold*, 109.
10. R. K. Maxwell, J. Thorkelson, L. L. Rogers, and R. B. Brander, "The Field Energetics of Winter-Dormant Black Bears (*Ursus americanus*) in Northeastern Minnesota," *Canadian Journal of Zoology* 66 (1988): 2095–2103.
11. R. J. Reiter, "Environmentally Determined Physiological Adjustments Made in Preparation for Hibernation," in H. C. Heller, ed., *Living in the Cold: Physiological and Biochemical Adaptations* (New York: Elsevier, 1986), 287–293.
12. Barnes et al., "Hibernators Periodically Arouse," 164.
13. C. P. Lyman, J. S. Willis, A. Malan, and L. C. H. Wang, *Hibernation and Torpor in Mammals and Birds* (New York: Academic Press, 1982), 45–49.
14. W. K. Milsom, "Metabolic Depression During Hibernation: The Role of Respiratory Acidosis," in Carey et al., *Life in the Cold*, 542.
15. Barnes and Ritter, "Patterns of Body Temperature Change," 119–130.
16. F. Strumwasser, "Some Physiological Principles Governing Hibernation in *Citellus beecheii*," *Bulletin of the Museum of Comparative Zoology* 124 (1960): 285–320.
17. C. P. Lyman et al., *Hibernation and Torpor*, 45.
18. M. Kawamichi and T. Kawamichi, "Factors Affecting Hibernation Commencement and Spring Emergence in Siberian Chipmunks (*Eutamias sibiricus*)," in Carey et al., *Life in the Cold*, 87.
19. Ibid., 88.
20. M. D. Tuttle, untitled cover note in *Bats* 13 (1)(1995), and M. Bruck Fenton, *Just Bats* (Toronto: University of Toronto Press, 1983), 73.
21. J. E. Hill and J. D. Smith, *Bats: A Natural History* (Austin: University of Texas Press, 1984), 145–147.
22. M. D. Tuttle, personal communication.
23. Hill and Smith, *Bats: A Natural History*, 146.
24. Ibid., 152.
25. Ibid., 151.

26. Unreferenced study cited by Hill and Smith, Ibid., 153.

27. D. R. Griffin, "Migrations and Homing of Bats," in W. A. Wimsatt, ed., *Biology of Bats* (New York: Academic Press, 1970), 233–264.

28. M. D. Tuttle, "Population Ecology of the Gray Bat (*Myotis grisescens*): Philopatry, Timing and Patterns of Movements, Weight Loss During Migration, and Seasonal Adaptive Strategies," *University of Kansas, Occasional Papers, Museum of Natural History* 54 (1976): 1–38.

29. Hill and Smith, *Bats: A Natural History*, 83.

30. French, "Patterns of Mammalian Hibernation," 573, and W. A. Wimsatt, "Some Interrelations of Reproduction and Hibernation in Mammals," *Symposium, Society for Experimental Biology* 23 (1969): 511–549.

Finding New Ground

1. G. R. Matlack, "Diaspore Size, Shape, and Fall Behavior in Wind-Dispersed Plant Species," *American Journal of Botany* 74 (1987): 1150–1160.

2. R. E. Farmer, Jr., *Seed Ecophysiology of Temperate and Boreal Zone Forest Trees* (Delray Beach, Fla.: St. Lucie Press, 1997), 47–54.

3. T. T. Kozlowski, *Seed Biology* (New York: Academic Press, 1972), Vol. I, 183.

4. Ibid., 184.

5. P. H. Raven, R. F. Evert, and S. E. Eichhorn, *Biology of Plants* (New York: Worth, 1986), 300–330.

6. Ibid., 614.

7. *The Writings of Henry David Thoreau* [Boston, 1906], Vol III, 17–20, September 24–25, 1851.

8. Farmer, *Seed Ecophysiology*, 59

9. M. Fenner, *Seed Ecology* (London: Chapman and Hall, 1985)

10. J. Burroughs, *Ways of Nature* (Boston: Houghton, Mifflin & Co., 1905), 252–253.

11. A. E. Sorensen, "Seed Dispersal by Adhesion," *Annual Review of Ecology and Systematics* 17 (1986): 443–463.

12. Ibid.

13. H. N. Ridley, *The Dispersal of Plants Throughout the World* (London: Reeve, 1930), cited in Sorensen, "Seed Dispersal by Adhesion," 451

14. B. W. Taylor, "An Example of Long Distance Dispersal," *Ecology* 35 (1954): 569–572.

15. C. Darwin, *The Origin of Species by Means of Natural Selection* (New York: D. Appleton and Company, 1914), Vol. II, 148.

Harvesting the Future

1. J. Burroughs, *Signs and Seasons* (Boston: Houghton Mifflin and Company, 1891), 57.

2. R. M. Lanner, *Made for Each Other* (New York: Oxford University Press, 1996), 22.

3. S. B. Vander Wall, *Food Hoarding in Animals* (Chicago: University of Chicago Press, 1990), 55–57.

4. S. B. Vander Wall and R. P. Balda, "Coadaptations of the Clark's Nutcracker and the Piñon Pine for Efficient Seed Harvest and Dispersal," *Ecological Monographs* 47 (1977): 89–111.

5. Ibid., 301, 303.

6. Ibid., 57.

7. Ibid., 181.

8. R. M. Lanner, *The Piñon Pine: A Natural and Cultural History* (Reno, Nev.: University of Nevada Press, 1981), 101–102.

9. Ibid., 48.

10. C. C. Abbott, *Days Out of Doors* (New York: D. Appleton and Co., 1889), 253–254

11. See various studies summarized by Vander Wall, *Food Hoarding in Animals* 156.

12. S. B. Vander Wall, "An Experimental Analysis of Cache Recovery in Clark's Nutcracker," *Animal Behavior* 30 (1982): 84–94.

13. Vander Wall, *Food Hoarding in Animals*, 166.

14. Ibid., 304, citing several observations by other researchers.

15. D. W. Macdonald, "Food Caching by Red Foxes and Some Other Carnivores," *Zeitschrift fuer Tierpsychologie* 42 (1976): 170–185.

16. Vander Wall, *Food Hoarding in Animals*, 174.

17. D. F. Tomback, "Dispersal of Whitebark Pine Seeds by Clark's Nutcracker: A Mutualism Hypothesis," *Journal of Animal Ecology* 51 (1982): 451–467, and S. B. Vander Wall, "Foraging of Clark's Nutcrackers on Rapidly Changing Pine Seed Resources," *Condor* 90 (1988): 621–631.

18. J. Bachman and J. J. Audubon, *The Quadrupeds of North America* (New York: Audubon Society, 1849), Vol. II.

19. J. Muir, *The Mountains of California* (Century Co., 1911), referring in this passage to a Sierra Nevada subspecies of the red squirrel known locally as the Douglas squirrel.

20. M. C. Smith, "Red Squirrel Responses to Spruce Cone Failure in Interior Alaska," *Journal of Wildlife Management* 32 (1968): 305–317.

21. C. C. Smith, "The Coevolution of Pine Squirrels (Tamiasciurus) and Conifers," *Ecological Monographs* 40 (1970): 349–371.

22. M. C. Smith, "Red Squirrel Responses."

23. S. B. Vander Wall, *Food Hoarding in Animals*, 244.

24. R. M. Lanner, *Made for Each Other* (New York: Oxford University Press, 1996), 31.

25. C. C. Smith, "Coevolution of Pine Squirrels."

26. S. B. Vander Wall, *Food Hoarding in Animals*.

27. B. G. Slough, "Beaver Food Cache Structure and Utilization," *Journal of Wildlife Management* 42 (1978): 644–646.

28. N. S. Novakowski, "The Winter Bioenergetics of a Beaver Population in Northern Latitudes," *Canadian Journal of Zoology* 45 (1967):1107–1118.

29. For a detailed review of winter adaptations in beaver and other semi-aquatic mammals, see P. Marchand, *Life in the Cold* (Hanover, N.H.: University Press of New England, 1996), 212–231.

Season of the Vole

1. S. D. West and H. T. Dublin, "Behavioral Strategies of Small Mammals Under Winter Conditions: Solitary or Social?," in J. F. Merritt, ed., *Winter Ecology of Small Mammals*, Carnegie Museum of Natural History Special Pub. 10 (Pittsburgh, 1984), 293–299.

2. J. O. Wolff and W. Z. Lidicker, Jr., "Communal Winter Nesting and Food Sharing in Taiga Voles," *Behavioral Ecology and Sociobiology* 9 (1981): 237–240.

3. J. Zejda, personal communication, 1984.
4. For a good summary of the processes associated with autumn weight loss see H. Hyvärinen, "Wintering Strategies of Voles and Shrews in Finland," pp. 139–148, W. B. Quay, "Winter Tissue Changes and Regulatory Mechanisms in Nonhibernating Small Mammals: A Survey and Evaluation of Adaptive and Non-adaptive Features," pp. 149–163, and V. A. Yaskin, "Seasonal Changes in Brain Morphology in Small Mammals," pp. 183–191, in Merritt, ed., *Winter Ecology of Small Mammals*.
5. D. D. Feist, "Metabolic and Thermogenic Adjustments in Winter Acclimatization of Subarctic Alaskan Red-Backed Voles," in Merritt, *Winter Ecology of Small Mammals*, 133.
6. D. C. Ure, "Autumn Mass Dynamics of Red-Backed Voles (*Clethrionomys gapperi*) in Colorado in Relation to Photoperiod Cues and Temperature," in Merritt, *Winter Ecology of Small Mammals*, 195.
7. J. O. Wolff, "Social Organization in the Taiga Vole (Microtus xanthognathus)," *Biologist* 62 (1980), 34–45.
8. E. Pulliainen and J. Keränen, "Composition and Function of Beard Lichen Stores Accumulated by Bank Voles, *Clethrionomys glareolus* Shreb.," *Aquilo Ser. Zoologica* 19 (1979): 73–76
9. P. M. Youngman, *Mammals of the Yukon Territory* (Otawa: National Museum of Natural Sciences, 1975).
10. Wolff and Lidicker, Jr., "Communal Winter Nesting and Food Sharing," 237–240.
11. A. N. Formozov, "Adaptive Modification of Behavior in Mammals of the Eurasian Steppes," *Journal of Mammalogy* 47 (1966): 208–223.
12. J. B. Benedict and A. D. Benedict, "Subnivean Root Caching by a Montane Vole (*Microtus montanus nanus*), Colorado Front Range, U.S.A." (in press).
13. Feist, "Metabolic and Thermogenic Adjustments," 135.
14. J. F. Merritt and D. A. Zegers, "Seasonal Thermogenesis and Body Mass Dynamics of *Clethrionomys gapperi*," *Canadian Journal of Zoology* 69 (1991): 2771–2777.
15. Feist, "Metabolic and Thermogenic Adjustments," 132.
16. Merritt and Zegers, "Seasonal Thermogenesis," 2771.
17. F. F. Mallory, J. R. Elliot, and R. J. Brooks, "Changes in Body Size in Fluctuating Populations of the Collared Lemming: Age and Photoperiod Influences," *Canadian Journal of Zoology* 59 (1981): 174–182.

Indian Summer

1. Quoted in C. Abbott, *Days Out of Doors* (New York: D. Appleton and Co., 1889), 280, evidently from the papers of William Hubbard, a minister in Ipswich, Massachusetts, during the mid 1600s.
2. R. Jefferies, *Nature Near London* (New York: Thomas Y. Crowell & Co., 1907), 173.
3. Ibid., 176.
4. Abbott, *Days Out of Doors*, 279–280.
5. W. Hubbard, *A Narrative of the Indian Wars in New-England: From the First Planting Thereof in the Year 1607, to the Year 1677* (Boston: John Foster, 1677).
6. J. Burroughs, *Winter Sunshine* (Boston: Houghton, Mifflin and Co., 1894).
7. E. W. Teale, *Autumn Across America* (New York: Dodd, Mead & Co., 1956), 188–189.

8. P. H. Gosse, *Romance of Natural History* (New York: A. L. Burt Company). The author's preface is dated 1860, Torquay, England. The front papers of the book carry the inscription "Registered at the Library of Congress November 1902 by New Amsterdam Book Company."
9. Abbott, *Days Out of Doors*, 279.
10. Ibid., 281.
11. Ibid., 282–283.
12. Z. Thompson, *Natural History of Vermont* (Burlington, Vt.: Zadock Thompson, 1853).
13. Ibid.

The Fall of the Leaf

1. R. Jefferies, *Field and Hedgerow* (London: Longmans, Green, and Co., 1900).
2. J. R. Bray and E. Gorham, "Litter Production in Forests of the World," *Advances in Ecological Research* 2 (1964): 101–157; and K. Van Cleve, L. Oliver, R. Schlentnee, L. A. Viereck, and C. T. Dryness, "Production and Nutrient Cycling in Taiga Forest Ecosystems," *Canadian Journal of Forest Research* 13 (1983): 747–766.
3. R. L. Donahue, R. W. Miller and J. C. Shickluna, *Soils: An Introduction to Soils and Plant Growth*, 4th ed. (Englewood Cliffs, N.J.: Prentice Hall, 1977), 155–157; R. L. Waring and W. H. Schlesinger, *Forest Ecosystems: Concepts and Management* (Orlando, Fla.: Academic Press, 1985), 184–186.
4. E. Schaechter, *In the Company of Mushrooms: A Biologist's Tale* (Cambridge, Mass.: Harvard University Press, 1997).
5. M. L. Smith, J. N. Bruhn, and J. P. Anderson, "The Fungus *Armilaria bulbosa* Is Among the Largest and Oldest Living Organisms," *Nature* 356 (1992): 428–431.
6. N. J. Dix and J. Webster, *Fungal Ecology* (London: Chapman and Hall, 1995), 97–98.
7. For an excellent review of the subject see G. C. Carroll, "Fungal Associates of Woody Plants as Insect Antagonists in Leaves in Stems," in P. Barbosa, V. A. Krischik, and C. G. Jones, eds., *Microbial Mediation of Plant-Herbivore Interactions* (New York: John Wiley and Sons, 1991), 253–271.
8. K. Clay, "Fungal Endophytes of Grasses: A Defensive Mutualism Between Plants and Fungi," *Ecology* 69 (1988): 10–16.
9. R. C. Cooke and A. D. M. Rayner, *Ecology of Saprophytic Fungi* (London: Longman, 1984).
10. Dix and Webster, *Fungal Ecology*, 44.
11. W. L. Mechaber, D. B. Marshall, R. A. Mechaber, R. T. Jobe, and F. S. Chew, "Mapping Leaf Surface Landscapes," *Proceedings of the National Academy of Science, USA* 93 (1996): 4600–4603.
12. Dix and Webster, *Fungal Ecology*, 94–96.
13. Ibid., 115.
14. M. J. Swift, "Basidiomycetes as Components of Forest Ecosystems," in J. C. Frankland, J. N. Hedger, and M. J. Swift, eds., *Decomposer Basidiomycetes: Their Biology and Ecology* (Cambridge: Cambridge University Press, 1982), 307–337.
15. Dix and Webster, *Fungal Ecology*, 125.
16. See data summarized from various sources in Waring and Schlesinger, *Forest Ecosys-*

tems, 197–201, and in S. H. Spurr and B. V. Barnes, *Forest Ecology* (New York: Wiley, 1980), 180–183.

17. J. R. Gosz, G. E. Likens, and F. H. Bormann, "Nutrient Content of Litter Fall on the Hubbard Brook Experimental Forest, New Hampshire," *Ecology* 53 (1972): 769–784, and J. D. Lousier and D. Parkinson, "Litter Decomposition in a Cool Temperate Deciduous Forest," *Canadian Journal of Botany* 54 (1976): 419–436.

18. From various studies summarized by Waring and Schleshinger, *Forest Ecosystems,* 174, 209.

19. Swift, "Basidiomycetes," 331–333.

20. Appearing in H. G. O. Blake, ed., *Autumn: From the Journal of Henry D. Thoreau* (Boston: Houghton, Mifflin and Company, 1892), 122.

21. Schaechter, *In the Company of Mushrooms,* 31.

22. Ibid., 36–37.

23. One of the more concise and easily readable summaries of meteorological data by season and geographic region appears in W. E. Reifsnyder, *Weathering the Wilderness* (San Francisco: Sierra Club Books, 1980).

Repose

1. H. Williamson, *The Lone Swallows* (London: W. Collins Sons, 1922), 179–180.

Index

Photo by Pedro Guerrero

Peter Marchand is a field biologist trained in both earth sciences and systems ecology at the University of New Hampshire. He earned his doctorate investigating physiological limits to tree growth at timberline and then devoted much of his career to the study of diverse forest, tundra, and, more recently, desert ecosystems, traveling widely throughout North America and Europe. His research interests often focus on winter phenomena; his book *Life in the Cold* (University Press of New England) is now in its third edition. But autumn, he says, has always been his favorite season for the fascinating changes taking place at this time of year. Currently a Visiting Professor at Colorado College and a frequent contributor to *Natural History,* Peter Marchand has conducted field research and shared his love of science with others for over twenty years. He calls Arizona home now ("when I'm not somewhere else"); there he ranges from low deserts to high mountain tops following his passion for the study of life at the extremes.